T0074231

John Manshreck
Transformation of the Electric Utility Business Model

The Global Energy Markets Series

Series Editor
Tom James

Volume 1

John Manshreck

Transformation of the Electric Utility Business Model

—

From Edison to Musk

DE GRUYTER

ISBN 978-3-11-071394-7
e-ISBN (PDF) 978-3-11-071403-6
e-ISBN (EPUB) 978-3-11-071412-8
ISSN 2629-3633

Library of Congress Control Number: 2021948220

Bibliographic information published by the Deutsche Nationalbibliothek
The Deutsche Nationalbibliothek lists this publication in the Deutsche Nationalbibliografie;
detailed bibliographic data are available on the internet at http://dnb.dnb.de.

© 2022 Walter de Gruyter GmbH, Berlin/Boston
Cover image: TeamDAF / iStock / Getty Images Plus
Typesetting: Integra Software Services Pvt. Ltd
Printing and binding: CPI books GmbH, Leck

www.degruyter.com

FSC
www.fsc.org

MIX
Papier aus verantwor-
tungsvollen Quellen
FSC® C083411

Acknowledgments

In many ways, writing a book would appear to be a solitary process. If that is the appearance, it is incorrect. This task would not have been possible without the support of many along the way.

The dozens of interview candidates who contributed to the research in this book did so with the understanding that their organizations and identities would be kept confidential, but must be thanked, nonetheless. I was consistently surprised by the generosity of so many who dedicated their knowledge and countless hours to the research. This could not have been completed without their contributions.

This book was formed out of research supporting a doctoral dissertation, and I was guided throughout the research process by Dr. Jonatan Pinkse. I discovered that Dr. Pinkse combines his expertise with an enthusiasm for his work that lightens the load of those working with him, and I have been extremely fortunate for his guidance. My thanks also to Dr. Vincent Mangematin and Dr. Valerie Sabatier who illuminated the concept of business models in the early stages of my research. And, in the later stages of the research Dr. Rene Bohnsack and Dr. Anne-Lorène Vernay invested great time and effort to guide development of the findings. Harold Nelson lent his deep operational experience to provide insight into challenging technical areas.

The book would never have been written without support from the publisher, as Jeffery Pepper convinced me that the subject was important and needed to get into print, while Jaya Dalal guided the task over the finish line.

Personal and business life has been affected by this. I have had many quizzical looks from business associates and friends when describing this quest. Through this, I have had nothing but support from Michael Weiss, my business partner of many years. And to my extended family, who patiently endured my excuses that I was too busy with writing to fulfill my family obligations, my thanks for your patience and support. And most of all, to my immediate family unit, Ann, Sarah, David and Katherine, who have provided unquestioning support throughout this process, my thanks.

And for those periods when writer's block reared its ugly head, my thanks to Yo-Yo Ma and Glenn Gould for the many times that their performances brought calm and focus back to the task at hand.

https://doi.org/10.1515/9783110714036-202

Foreword

Till recently, nobody in their wildest dreams would have imagined an Electric Vehicle (EV) in space! Current events have shown us the unusual and unexpected coming out of an enthusiastic genius, Elon Musk. What Edison and Tesla started, Musk took to new heights, literally!

Transformation of the Electric Utility Business Model by John Manshreck gives power packed, electrifying information all decked up in twelve chapters/topics which can be best described as concise yet detailed. I found the writing of this foreword worthy and exciting as this is an important and timely topic to be covered as the world focusses on renewable and clean energy and how to deliver that to the consumer. This is a timely and valuable contribution to the GEMS Global Energy Market Series which aims to bring forth valuable written contributions on topics related to the decarbonization of the economy and future approaches to provide affordable clean energy for all.

Whilst reading through this book I was really impressed by the way John Manshreck has presented minute yet interesting details about the changes and transformation it took for electric utility and mobility to reach where we are today.

The chapters covering the business model, risks involved, emergence of technology and the current need for infrastructure to boost the energy industry make for good reading for any entrepreneurs out there looking to optimize the industry tools of the future and enhance the way things work to make earth a sustainable place to live going forward.

A quote from the book, very well said by Elon Musk, highlights the crux of the biggest problem we might face in the coming years in the electric business: scalability of solutions. Musk said, "If you have a great solar roof, and you have a battery pack in your house, and you have an electric car, that scales worldwide. You can solve the whole energy equation with that." The answer though perhaps lies in lots of small solutions solving the bigger problem... as he explained in these three simple lines.

The tables/graphs are representative of the superb information throughout this book. It is packed with detailed and easily accessible useful information.

The book also guides individuals willing to enter the industry about all the competition they might face and where they can find opportunities to build business models that can stand up against strong incumbent industry players. The intense research done in writing this book can be gauged by the number of people consulted and industries visited.

Transformation of the Electric Utility Business Model contains a plethora of right information at the right place and should be on the desk of all those who wish to transform the industry while keeping the utmost priority of climate change that nations around the world are pledging. This book is essential reading for all those

https://doi.org/10.1515/9783110714036-203

willing to grow their business keeping in mind the forecast of emerging business needs and industry norms.

Tom James

CEO / CIO – Cofounder

TradeFlow Capital Management Pte Ltd.

Contents

Part 2: **Utility Business Model Transformation – A History**

Part 3: **Emergence of a New Business Model**

Introduction

The development of the electrical grid has been an incredible feat, described by the American Academy of Sciences as the greatest engineering achievement of the twentieth century (Constable & Somerville, 2003). It is remarkable that this quiet technology that we so seldom think about, should be regarded as a greater feat of engineering than the automobile, the airplane, the computer, or the internet. Today's electrical grid, described as "the worlds largest machine" (Howe, 2016), has a lineage that traces back over a century to some of the greatest inventors and entrepreneurs of the late nineteenth century, with founders' names that still resonate today, such as Siemens, Ferranti, and Thomson in Europe, and Westinghouse, Tesla, and Edison in North America.

From these great entrepreneurs, the grid has expanded across the developed world to enable a century of economic growth. It has embedded itself into our lives to an extent that billions of people regard the availability of electricity at the flick of a switch as an essential service. However, there is increasing evidence to show that this great achievement of the twentieth century is straining to meet the demands of this century. Not only are many parts of the grid straining in a technological sense, but the business model of the utilities that operate the grid are under increasing stress. For over fifty years following the Great Depression, the utility business model was so robust that not a single large public utility in North America declared insolvency (Berry, 1988). In the last two decades, however, the sector has seen some of the largest bankruptcies in American history, with the insolvency of storied names like Calpine, Pacific Gas and Electric (which actually went bankrupt twice, in 2001 and 2019) and, of course, Enron. In Europe, utilities have also struggled financially. In Germany, the country's four largest utilities were among the country's fifty largest companies in 2016. But only two years later, after restructuring to support the "Energiewende," Germany's transition to a nuclear-free, low carbon economy, the utilities failed to list among the country's one hundred largest companies (Clean Energy Wire, 2018). Utilities and their business models around the world are under strain.

After decades in the utility sector, while seeing evidence of the stress and change in many parts of the industry, I started to become aware that the industry was at the center of one of the dominant issues of our time: decarbonization of the economy. At the same time utilities were struggling with risk management and resilience of the energy system, as increasingly frequent extreme weather events would leave hundreds of thousands of customers without electricity for days. I wanted to understand what was happening to utilities attempting to respond to these challenges, so I spent several years in doctoral studies and research (with a fantastic academic supervisor and some wonderfully cooperative utilities), conducting case study research of North American utilities and the business model changes that they were facing. The result was thousands of pages of interview notes and analysis, a doctoral dissertation, and

https://doi.org/10.1515/9783110714036-205

ultimately the publication of this book. The research that the book is based upon uses the American experience more than from any other country, but the story it tells was never intended to be solely an American story. With focus on the effect of the broad issues of resilience and decarbonization on the utility business model, rather than deep analysis of issues specific to a particular jurisdiction, the story is intended to be relevant to a broader international audience.

This book is intended to appeal to several types of readers. First, the book will appeal to managers, policy makers, and regulators who need to understand the electric sector and its central role in issues destined to be at the center of public debate over the coming decade: energy security and resilience, and decarbonization. Second, the book is intended to appeal to teachers and students, particularly those in fields related to sustainability, engineering, resource economics, or energy studies, who wish to understand the history of the evolution of the electric utility sector, and the issues it currently faces. It will also appeal to students of business strategy who wish to understand business models, how business models become dominant, and how those models transition over time from one dominant model to the next. Finally, it is intended for general geeks with a desire to understand sustainability and business strategy. I count myself in the latter category.

Many studies of the future of electric utilities tend to look deeply at a single aspect of the sector, such as new generation technologies, regulatory issues, or the impact of new requirements from customers deploying products such as electric vehicles. Since this book deals with the history and the current stresses on the utility from the perspective of its business model, it takes a somewhat broader approach. The book will focus on integration of broad knowledge of many areas into a business model framework rather than the deep investigation of a particular technology or issue.

This book has three parts (see Figure 1). Part 1 contains two chapters, each providing a foundation used in the rest of the book. The first chapter describes an environment that is dominated by two key issues that have become critical for utility management over the past few years: the management of the risk of infrequent but catastrophic events, and the imperative for utilities to support a decarbonized economy. The second chapter outlines for the reader the structure of the "business model" framework, a concept that forms the basis for analysis and discussions throughout the book.

In the belief that one does not understand the present without first understanding the past, Part 2 of the book describes the history of the electric utility business model. The story is told in four chapters (chapters 3 to 6). Chapter 3 describes the business model used in the 1800s to produce electricity before the arrival of the first electric utility, which in certain aspects predicts today's emerging technologies that allow customer self-generation of electricity. Chapter 4 describes Thomas Edison's ground-breaking business model that supported the introduction of the first electric utility in 1882. Chapter 5 describes the ascendency of a new utility business model

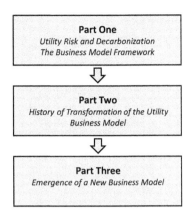

Figure 1: The Main Elements of the Book.

in the 1890s based on alternating current (AC) technology, which allowed incredible growth and cost reduction, but resulted in de facto monopoly power by a single utility in many jurisdictions, and inefficient competition in others. Chapter 6 describes the third utility business model, introduced in the 1920s and 1930s, in which utilities were provided a local or regional monopoly in exchange for an obligation to serve all customers within that region, and a regulated rate of return. A century after its introduction, this still forms the basis of today's dominant utility business model.

Part 3 of the book describes the current and possible future state of the utility business model and the key challenges it will face over the next couple of decades. The story is told in six chapters (chapters 7 to 12). Chapter 7 describes the key issues disrupting the current utility business model. Chapters 8 to 11 describe the stresses placed upon each of four major components of the business model: customer identification, the value offering, the value chain, and the utility's means of value capture. Chapter 12 describes the possible shape of the new and still evolving utility business model.

Those readers who wish to focus on current issues may wish to focus solely on Parts 1 and 3. However, Part 2 does provide an understanding of how the utility business model arrived at its current state and will help the reader better understand some of the current roadblocks and path dependencies that impede changes to the business model. (Also, it is quite interesting.)

Part 1: **Utilities at Risk**

The electrical grid is part of everyday life around the world, with billions of people dependent upon it for everything from light, heat, communication, and, increasingly in the coming decades, transportation. However, much of the grid is decades old, and the fragility of the system is sometimes exposed by a simple event. In June 2019, for example, a short circuit on a transmission line, probably caused by lightning or a fallen tree branch, caused 48 million people in Argentina, Uruguay, and parts of Paraguay to be without power for up to fourteen hours (Henao & Byrne, 2019).

The inherent fragility of the system is not a new observation. In a study commissioned by the US Pentagon forty years ago, Amory and L. Hunter Lovins warned about the vulnerability of the electrical system:

> energy vulnerability is an unintended side effect of the nature and organization of highly centralized technologies. Complex energy devices were built and linked together one by one without considering how vulnerable a system this process was creating.
>
> (Lovins & Lovins, 1982, p. 2)

The report concluded that the inherent brittleness of the electrical system posed a greater threat to American national security than the risk of the disruption of imported oil (Bakke, 2016), a remarkable conclusion to come so soon after the 1970s oil crisis. Despite the warning, in the following decades the grid has continued to be dominated by a centralized architecture, with utility business models reflective of that centralization. Although recent decarbonization efforts in many jurisdictions have resulted in growing levels of decentralized renewable generation, today's utilities manage risk and decarbonization issues from a business model still adapted to the centralized architecture of the last century.

Chapter 1 discusses two issues that have recently become central priorities for utility management: first, the management of the risk of catastrophic events (referred to as "black sky risk"), and second, the management of this risk in an era of decarbonization. These are both long-term issues that will take many years to address, and in other times might slip to the bottom of a management agenda normally focused on short-term topics. However, recent events have pushed the management of black sky risk and decarbonization to the very top of utility priorities and clarify the urgent need for a grid that is both resilient and decarbonized. Although the management of black sky risk and decarbonization are connected and best managed jointly, indications are that the management of these issues can be constrained by current utility business models.

Chapter 2 outlines for the reader the structure of the "business model" framework, a management tool that has grown in prominence in the past two decades. The components of the business model framework form the basis for analysis and discussions throughout the book. The business model that is currently dominant in the utility sector will be shown in later chapters of the book to pose a potential impediment in transitioning to a resilient and decarbonized grid.

https://doi.org/10.1515/9783110714036-001

1 Twin Challenges: Risk Management and Decarbonization

> The next few years are probably the most important in our history.
> Dr. Debra Roberts, co-chair, Intergovernmental Panel on Climate Change (IPCC, 2018a, p. 2)

Today's utilities are confronting the risk of high impact events that in the past were considered so improbable as to be outside most planning scenarios. In February 2021, for example, an unusually cold stretch of weather drove temperatures below freezing in Texas, and soaring demand for electricity forced grid operators to implement a series of rolling blackouts. These blackouts turned into outages that lasted for days, as the grid failed to respond to the strain placed upon it, and more than four million homes and businesses in the state were left in the cold and dark without electricity (Canizales, 2021). Without electricity, water treatment plants were unable to operate, leaving thirteen million people without safe drinking water. Texas, which would be the world's tenth largest economy if measured as a sovereign nation, had dozens of its residents die from failed infrastructure, many of hypothermia in their own homes (Mulcahy, 2021). The duration and impact of these blackouts was very nearly much worse. In the early morning hours of February 15, according to the Texas grid operator, the state's electrical system was minutes away from a total failure that could have left Texas without electricity for weeks instead of days (Jenkins, 2021).

Only weeks before these storms were leaving millions without power in Texas, the new White House administration was issuing an executive order to direct the US electricity industry to reduce net greenhouse gas (GHG) emissions to zero by 2035, and to support the economy to achieve net-zero emissions by 2050 (*The Economist*, 2021). In an industry that often measures the life of its assets in multiple decades, the means of producing and distributing electricity throughout the country is to be restructured in less than a decade and a half.

These two events were not directly related, but they pointed to two long-term issues that will dominate the agendas of utility management and regulators in the coming years: managing the risk of catastrophic events, and decarbonization. Furthermore, the two events could have just as likely occurred in Europe, Asia, Australia, or any developed economy. The electrical utilities in these economies all face challenges of managing the risk of catastrophic events, while rebuilding their electrical infrastructure to meet tightening emission standards. Fortunately, with coordinated planning these objectives are not incompatible. In fact, many investments in system resilience can support decarbonization, and vice versa. However, as discussed later in this book, regulators and management will find that many of the envisioned investments are not compatible with current utility business models and will require creativity and new regulatory frameworks to successfully implement.

https://doi.org/10.1515/9783110714036-002

Black Sky Risk

"Black swan risk," is a phrase that became widely used in financial markets in the late 2000s as economists and analysts struggled to understand the global financial crisis that rocked stock and bond markets around the world. Although the phrase had been used for centuries, it had been recently popularized in a 2007 book by Nicholas Taleb to describe a low-probability, high-impact event that is very difficult to forecast, but in hindsight would appear to be predictable (Taleb, 2007). The term "black swan" had been used in London in the 1600s to describe a circumstance that was thought to be impossible. Up to that time, all observations of swans in Europe had confirmed that swans were white and, based on the evidence, the existence of a black swan was assumed to be impossible. This was a belief that continued until a Dutch expedition to Australia in 1697 discovered, much to the surprise of the Dutch sailors, the existence of black swans. This example was developed in philosophy to describe the "black swan fallacy," a fallacy of assuming that if something has never been witnessed, it must be impossible (Taleb, 2007). The concept was updated by Taleb in 2007 to describe a black swan as an event that has three attributes. First, because there is no history that would indicate it to be a possibility, the event lies outside the boundaries of normal expectations. Second, the event is highly impactful. Third, people tend to develop explanations for the event in hindsight, to make the event appear explainable and predictable (Taleb, 2007).

Like financial markets, the utility sector in recent years has also focused attention on the risk of extremely adverse events (sometimes, unfortunately, in reaction to the occurrence of such events). The term developed within the utility sector is "black sky risk," used to describe a low probability event in a utility environment that carries catastrophic consequences and disrupts normal functioning of critical infrastructure over a wide area, for an extended duration (EIS Council, 2021). The risk of black sky hazards can arise from two main sources: natural factors and man-made malicious actions (see Table 1.1). Natural factors include events such as earthquakes or wildfires, and events related to extreme weather, such as ice storms or heat waves. Natural events can also include "geomagnetic disturbances" caused by solar flares, which are relatively rare but have the potential to be very impactful since they can damage key distribution assets, such as transformers,[1] over a wide area (NASEM, 2017).

Man-made malicious actions include rare events such as direct terrorist attacks on facilities or the growing threat of cyberterrorism. Developed nations are fortunate that the impact of such attacks has been limited. However, the risks posed by such attacks, and cyberattacks in particular, are very real and could cause major disruptions to the electric grid (NASEM, 2017). *The New York Times* has reported, for example, that in the early years of the Obama administration, the United States had

---

1 See the glossary for a definition of "transformer."

Table 1.1: Examples of High Impact, Low Probability Events.

Description	Example
Man-Made Malicious Actions	
Cyberattack	Eastern Ukraine (December 2015): Cyber intruders launched a cyberattack on the utility's system control center, impairing the utility's response capability. The attackers then took control of substations and shut down power from those substations. The attack affected only 225,000 people for about six hours, but more importantly, demonstrated a capability to create much greater disruption over a larger area (NASEM, 2017).
Natural Event	
Geomagnetic Disturbance	Eastern Canada (March 1989): Solar flares created a magnetic storm resulting in the loss of five transmission lines in Eastern Canada and outages affecting approximately six million people (NASEM, 2017).
Hurricanes	North Eastern United States (October 2012): Hurricane Sandy hit New York and New Jersey and neighboring states, resulting in outages for nearly 5.8 million customers for up to ten days (NASEM, 2017).
Ice Storm	Quebec and New York (January 1998): A series of storms with freezing rain resulted in the collapse of over seven hundred transmission towers. The cascading failure left more than five million people without power, and the response required the replacement of 1,800 miles (about 2,900 km) of transmission and distribution circuits, 26,000 poles, and 4,000 pole-top transformers. Three weeks after the storm, hundreds of thousands of customers were still without power (NASEM, 2017).
Wildfires	California (October 2019): The risk of wildfires during periods of heat and wind caused California utilities to intentionally de-energize transmission and distribution lines, resulting in a series of outages to 2.8 million customers over five days (Hay, 2019).
Cold Weather	Texas (January 2021): The rolling blackouts that grid operators implemented to manage soaring demand turned into outages that lasted for days. With the lack of electricity, water treatment plants were without power, leaving entire cities without safe drinking water (Canizales, 2021). Dozens of people died, many of hypothermia in their own homes (Mulcahy, 2021).

developed a plan, code-named "Nitro Zeus," to conduct a cyberattack that would disable the Iranian electrical grid in the event of conflict escalation, bringing total disruption to Iranian infrastructure without firing a shot (Sanger & Mazzetti, 2016). The United States is certainly not the only country to aspire to this capability. Man-made actions could also take the form of an electromagnetic pulse (EMP) arising from the detonation of a nuclear weapon at high altitude. Like a solar flare, an electromagnetic pulse can damage transmission and distribution hardware, such as transformers and circuit breakers, over a wide area. However, it can also disable digital equipment like computers, sensors, and control devices that are not hardened to military specifications (NASEM, 2017). Military organizations have also developed the capability to deliver a non-nuclear electromagnetic pulse from a device the size of a suitcase, to disable local power equipment in locations like substations or power plants (NASEM, 2017).

The First Objective: Managing Black Sky Risk

As a first step in managing black sky risk, it is necessary to understand the difference between reliability and resilience, and that each requires a different management approach.

Reliability management is a well-established discipline in the utility sector. The disciple utilizes metrics based on many data points gathered from years of day-to-day operations, to forecast patterns of frequent events that affect reliability. For example, utilities have tens of thousands of transformers, and measure the probability that a transformer of a particular age will perform without failure each year. These measures will then be used to plan maintenance and scheduled replacement of the transformer fleet. This, with other maintenance programs, will affect the utility's measures of reliability, such as the frequency and duration of power outages. These metrics provide a basis to measure a utility's performance over time, and a comparison against industry benchmarks for reliability.

Managing the risk of black sky events requires a much different approach than reliability management. Reliability management can use abundant data to focus on reducing the probability of an event, such as equipment failure. The frequency of a black sky event, such as an earthquake or hurricane, however, is so low, and the data set of occurrences so small, that metrics are much less useful in guiding a utility's preparations for such events. Instead, management of black sky risk focuses on reducing the consequence of the event by building system resilience (Wender, Morgan, & Holmes, 2017). The difference was described by the American National Academies of Sciences, Engineering, and Medicine:

> Resilience is not the same as reliability. While minimizing the likelihood of large-area, long-duration outages is important, a resilient system is one that acknowledges that such outages can occur, prepares to deal with them, minimizes their impact when they occur, is able to restore service quickly, and draws lessons from the experience to improve performance in the future.
> (NASEM, 2017, p. 10)

Unlike reliability, the resilience of a utility is difficult to measure (Wender et al., 2017). With the absence of resilience metrics, it is difficult for a utility to compare alternative investment strategies to achieve resilience, to measure resilience against peers, or to assess changes in resilience over time. As a result, business cases to justify investments to manage black sky risk often identify resilience as an unquantified, intangible benefit (Rickerson, Gillis, & Bulkeley, 2019).

While investments in resilience have often been reactive, utilities and their regulators are starting to adopt a more proactive planning approach to the subject. The management of resilience recognizes that the risk of a black sky event can not be eliminated, but with planning and proactive measures, the consequence of its occurrence can be managed and mitigated (NASEM, 2017). The range of activities available to utilities to manage this risk can be broken into five main categories: system hardening, operational response, resilient design, system decentralization, and transmission infrastructure.

System Hardening

The goal of system hardening is to reduce black sky risk by making the existing system less vulnerable to that risk. This can include investment in the electrical system to make it less susceptible to physical damage, through measures such as undergrounding of transmission and distribution lines, structurally reinforcing towers and poles to withstand greater physical stress, or siting facilities where they are less susceptible to natural disaster, such as flooding from storm surges (NASEM, 2017). A system can also be hardened with "self-healing grid" technology, which uses sensors, automated control devices, and outage management software to automatically respond to problems (Kwasinski, Andrade, Castro-Sitiriche, & O'Neill-Carrillo, 2019). These technologies can operate with little or no human intervention to quickly isolate faults and restore service to remaining customers.

Utilities can harden their system against the risk of cyberattack through continual investment in monitoring and control systems, and through coordinated sharing of information at an industry level and through national defence agencies. The task of shielding from a malicious electromagnetic pulse attack will generally be coordinated with national defence agencies (NASEM, 2017).

Operational Response

Once the need to respond to a black sky threat has been identified, the utility needs to identify shortcomings in its response capability and invest in upgrading that capability. This will include investment in people, processes, and coordination tools to guide responses, including simulation training and coordination with external disaster response organizations (NASEM, 2017). Activities would also include assessment of spare inventory for key infrastructure components. For example, substations contain equipment like high-voltage transformers that are often customized for a specific location, and have long procurement times (NASEM, 2017). The maintenance of redundant supplies for such key elements can reduce restoration time following a black sky event by days or even weeks.

Resilient Design

Most electrical distribution systems have been traditionally configured as a "radial" network, similar in appearance to a tree, with generation feeding into the trunk of the tree and customers spread along the branches (Prakash, Lallu, Islam, & Mamun, 2016). Although relatively simple and inexpensive to design and construct, an unfortunate characteristic of the radial design is that an outage at any point on a branch impacts all customers further down the branch (Prakash et al., 2016). An alternative to this design is a networked architecture in which circuits are interconnected, and capable of being quickly reconfigured by distribution automation technology to mitigate service interruptions. Although best suited for densely populated urban areas, and more expensive than a traditional radial system, a networked architecture increases system resilience and reliability and more easily accommodates multiple service connections to critical customers like hospitals (Prakash et al., 2016). In addition to reducing the number of customers affected by a black sky event (NASEM, 2017), a networked architecture can also accommodate higher levels of distributed renewables (Prakash et al., 2016).

System Decentralization

One of the most significant changes in the nature of the utility sector in the past two decades has been the growth of distributed energy resources (DERs). (Distributed energy resources will be discussed in detail in chapters 7 and 8. See the glossary for a short description of DERs.) These are energy resources that are typically located on a customer's premise and supply all or part the customer's electric load and may feed surplus generation into the grid. Examples of decentralized resources include solar photovoltaics, diesel generators, small natural gas turbines, independent battery storage (such as Elon Musk's Tesla Powerwall) and, potentially, the batteries of a growing fleet of electric vehicles. With coordinated planning and upgrading of the electrical distribution system, and the development of monitoring and control capability, these distributed energy resources have the potential to act as a grid resource at the end of vulnerable distribution lines (NASEM, 2017). The distributed resources could partially energize local distribution to service critical local loads, even if the larger electrical system is damaged by a black sky event (NASEM, 2017). A decentralized system can also include microgrids,[2] which are localized grids with their own generation source that can disconnect from the traditional grid to operate autonomously. Since microgrids are able to continue to operate while the main grid is down, microgrids can strengthen grid resilience and mitigate grid disturbances.

---

2 See the glossary for a definition of "microgrid."

Transmission Infrastructure

Although there were many issues that contributed to the severity of the Texas blackouts of February 2021, many industry observers noted that an increase in interconnections between Texas and other grids could have offset some of the generation capacity lost when the cold weather hit the state (Trabish, 2021). Texas is unique in the continental United States in choosing to operate an energy grid that does not interconnect with the transmission systems of neighboring jurisdictions (Trabish, 2021). Although this ensures that its energy system is independent and free from federal oversight, it also means that Texas can not draw from its neighbors during critical events. The few parts of the state that are not part of the main Texas grid, such as El Paso, but connected instead to the larger grid that connects to other western states, avoided the rolling blackouts that affected the rest of the state in February 2021 (Rust & Kim, 2021).

The interconnection of transmission infrastructures across a large area allows regional grids to access a greater range of generation sources and provides a more flexible supply of generation in the face of a supply disruption (Roberson et al., 2019). This was demonstrated in the North American "polar vortex" cold wave of January 2019, when transmission lines between regional grids delivered power from the eastern grids of North America to a frozen mid-west, and days later reversed flow to supply eastern grids as the cold wave moved to the east coast (Gramlich, 2021). These transmission interconnections will only increase in importance with the need to transport new sources of utility scale renewable energy from wind and solar farms in remote regions to centers of customer load.

The Second Objective: Designing for Decarbonization

Electrical distribution utilities sit at the center of what is, arguably, the most pressing issue of our times: global warming. The October 2018 report of the Intergovernmental Panel on Climate Change (IPCC) set out the actions required to limit an increase in global temperatures to 1.5 degrees Celsius (IPCC, 2018b). The report estimated this target would require 75% of the world's electricity to be sourced from renewables by 2050, an increase from just 23% in 2015 (Merchant, 2018). Since that report, the rapid decline in costs of renewables and new political agendas have energized governments to exceed even these targets. In January 2021, for example, the new American president signed an executive order requiring the US electricity sector to reduce net GHG emissions to zero by 2035, and to support the economy to achieve net-zero emissions by 2050 (The Economist, 2021). Other countries are similarly revisiting their own programs for GHG reductions.

Utilities are central to this massive decarbonization effort, and are being called upon to enable, integrate, and manage an unprecedented investment in renewable resources on both sides of the customer meter.[3] Utilities are also being called upon to enable the decarbonization of other sectors of the economy, such as transportation, while also increasing their system resilience. However, many of the actions required for decarbonization and resilience do not fit existing utility business models, and regulators and utilities will be called upon to reconfigure regulatory frameworks to support these efforts.

The following section will briefly examine the misalignment of aspects of the traditional utility business model with current decarbonization and resilience efforts. The broader impact of decarbonization as a disruptor within the utility sector is discussed in Chapter 7.

The Good News: Resilience and Decarbonization Compatibility

Fortunately, with planning, many of the investments required to make a system more resilient will also support the decarbonization of the electrical system. This is reflected, for example, in the recent proposals for redevelopment of Puerto Rico's electrical grid.

In September 2017, the Caribbean island of Puerto Rico was hit by hurricanes Irma and Maria only days apart, causing more than $100 billion in damage and killing almost three thousand people (Coto, 2020). Almost all transmission infrastructure in the eastern half of the island was severely damaged, 74% of substations were damaged, and about one tenth of the island's distribution poles had to be replaced (Kwasinski et al., 2019). Power was lost to essentially all the island's 3.4 million residents, and electrical service was not fully restored through the island for over ten months (Kwasinski et al., 2019). In planning the rebuild of the island's energy system, the US Department of Energy commissioned experts to provide technically sound recommendations for Puerto Rico's energy investment decisions, with a particular focus on resilience and decarbonization of the system (Narang et al., 2021). These studies lay out an investment path that includes enhanced transmission infrastructure to integrate utility-scale renewable electrical generation, grid level storage, microgrids, and streamlined connections to a growing fleet of distributed resources, particularly solar photovoltaics. In addition to adding to the electrical system's resilience, these actions also enable a substantial increase in the share of Puerto Rico's generation from renewable resources, targeting a goal of 40% renewables by 2025 and 100% by 2050 (Narang et al., 2021). The recommendations address the twin challenges of resilience and decarbonization

---

3 As discussed later in the book, the customer meter is the traditional line of demarcation for the ownership and regulation of utility assets, with the utility traditionally owning assets up to and including the meter, and the customer owning assets on their side of the meter.

faced by so many utilities and demonstrate that the two issues are not incompatible when pursued simultaneously.

Earlier in this chapter, we identified five main types of investment that increase system resilience: system hardening, operational response, resilient design, system decentralization, and transmission infrastructure. Of these, two classes of resilience-based investments are particularly important to the decarbonization of the grid: system decentralization and transmission infrastructure.

System Decentralization

The expert report on the reconstruction of Puerto Rico's electrical grid proposed a substantial investment in distributed energy resources, including rooftop solar photovoltaics, battery storage and microgrids (Narang et al., 2021). This type of investment represents a fundamental shift from a traditional utility structure built on centralized generation, bulk transport of the electricity over a transmission system, and final delivery to the customer through a distribution system. Instead, distributed resources are naturally located near the customer, and if properly configured they can continue to provide service despite disruptions elsewhere in the electrical system.

The practice of including distributed resources into resilience planning is still new. One of the reasons for their past absence is that a quick cost calculation could quickly rule out renewable distributed resources as an option in many plans. Today, the cost of a combined solar and battery system is still expensive[4] for many businesses and homeowners, and not yet competitive in many jurisdictions with the cost of traditional electricity from the grid combined with a gasoline backup generator. However, while the cost of grid supplied electricity has been flat or rising in most jurisdictions, the cost of renewable alternatives has been falling sharply. The benchmark cost of an installed residential solar photovoltaic system has fallen by half in the past decade (Feldman, Ramasamy, Fu, Ramdas, Desai, & Margolis, 2021) (see Chapter 7, Figure 7.1), while battery costs have declined by 89% (see Chapter 7, Figure 7.3), driven by scale economies and advances in lithium-ion battery chemistry (BloombergNEF, 2020a). The trend is forecast to continue. As Elon Musk, the CEO of Tesla predicted at the end of 2020, the cost of batteries supplied by Tesla would be cut by a further 56% over the following two years (Templeton, 2020).

The resilience offered by batteries can be of particular value to critical customers, like police stations or firehalls, that have traditionally relied upon diesel generators for

---

4 According to the US National Renewable Energy Lab, the benchmark cost in 2020 of a combined residential PV solar and battery combination was about $19 thousand for a typical 7 kw rooftop PV system, plus a further $17 thousand for an installed 5 kw, 20 kwh battery system (such as a Tesla Powerwall), for a combined installed cost of about $36 thousand (Feldman, Ramasamy, Fu, Ramdas, Desai, & Margolis, 2021).

backup. Unlike a back-up diesel generator that is operational only during a grid outage and can only supply electricity for as long as it has fuel, a backup system based upon customer sited renewable solar and battery technology offers the advantages of the being able to refuel when the grid is out of service. Although not yet widely deployed to service large facilities, the declining cost of customer-sited renewable generation and energy storage is rapidly increasing its attractiveness relative to gasoline or diesel generators (Rickerson et al., 2019). However, successful deployment at any facility requires a design that includes adequate battery and inverter[5] capacity, appropriate sizing of solar panels to recharge the batteries, and limitation of electrical devices on the circuits backed up (Cinnamon, 2019). (Separation of electrical circuits servicing critical loads, such as refrigeration, from non-critical loads is typical for any backup installation, not just for those based on solar and batteries.)

Batteries are not only being considered for resilience at customer premises. Utilities are also taking advantage of falling costs to install large batteries at critical junctions on the grid to increase both efficiency and resilience on the grid. As renewable generation becomes more dominant with its pattern of intermittency, large battery storage facilities on the grid are being built to smooth supply and demand, while adding resilience when parts of the system go down. In Australia, for example, the 300 MW/450MWh "Victorian Big Battery" utilizing Tesla Megapack batteries, will be one of the world's largest when complete in 2021 and able to power half a million homes for an hour (Morton, 2020). However, it will be soon overtaken by larger projects planned in Germany, Saudi Arabia, the United States, and others on the drawing board.

Batteries from electric vehicles also offer intriguing promise for resilience at the local level. With a typical household in the United States consuming about 28 kWh per day (US EIA, 2020c), a fully charged Chevy Bolt or Tesla Model 3, equipped with a battery sized at about 60 to 70 kWh, has the theoretical potential to supply that household for more than two days from a fully charged battery. In Japan, Nissan is working with local governments and utilities to make batteries of the Nissan Leaf available to energize homes and businesses in emergency. In September 2019, as Japan was hit by Typhoon Faxai, Nissan sent more than fifty of its Leaf electric vehicles to provide backup power to community centers east of Tokyo (Gerdes, 2019). The integration of electric vehicle batteries into this type of function is years away from mainstream adoption and will require investment in inverters and controls to energize circuits in a designated facility, but as costs decline and EV deployment widens it offers potential as a means of increasing system resilience.

---

5 See the glossary for definition of "inverter".

Transmission Infrastructure

In planning the reconstruction of Puerto Rico's grid, the expert report commissioned by the US Department of Energy confirmed that an enhanced transmission infrastructure is foundational to resilience (Narang et al., 2021). This is consistent with conclusions found in jurisdictions around the world, as the need to maintain system resilience while shifting to renewable generation is driving a general demand for increased transmission capacity. The increase in transmission capacity can be met not only by building new transmission lines, which can be fraught with political and environmental opposition, but also by investing in advanced technologies that can increase capacity of existing lines. The application of these technologies can reduce congestion on existing transmission lines and increase line capacity, often at less cost than the construction of new lines (Tsuchida & Gramlich, 2019). For example, through investment in sensors and control systems, the state of a transmission line can be measured in real time. This capability, known as "Dynamic Line Rating (DLR)," allows the system operator to operate the transmission system more efficiently, reducing grid congestion, and effectively increasing the capacity of the existing system (Tsuchida & Gramlich, 2019).

The value of increased transmission capacity to enable resilience and increased renewable resources arises from several factors. First, large-scale renewable resources, such as offshore wind farms or utility scale solar farms, tend to be in the windiest and sunniest parts of the country and are rarely located near load centers. Transmission systems are needed to move this clean generation to centers of customer load (Larson et al., 2020). Second, transmission capacity can increase the diversity of the portfolio of renewable resources available to a jurisdiction in the event of an adverse "black sky" event. As the people of Texas discovered in their outage of early 2021, access to a wider portfolio of generation resources might have mitigated the impact of the loss of their own generation. Third, renewable resources like solar and wind, are by their very nature, as variable as the weather. However, as the size of a geographic area expands, the average variability of the weather declines (MacDonald, Clack, Alexander, Dunbar, Wilczak, & Xie, 2016). If the wind is not blowing or the sun not shining in one region, the probability that they are available elsewhere will increase as the size of the geographic area increases. Using transmission capacity to connect larger areas that are rich in solar and wind resources will, generally, increase both the reliability and efficiency of the total system (MacDonald et al., 2016).

This increase in transmission capacity will require substantial investment over the next several decades. For example, a 2020 study by researchers at Princeton University examined various scenarios that could lead to a decarbonization of the American economy by 2050. The researchers found that, even under the scenario with the least expansion of transmission capacity, the country's transmission systems will still need to grow by 60% by 2030, and to triple by 2050, with capital investment of $360 billion by 2030 and $2.4 trillion by 2050 (Larson et al., 2020). As one of the study's authors noted:

The current power grid took 150 years to build. Now, to get to net-zero emissions by 2050, we have to build that amount of transmission again in the next 15 years and then build that much more again in the 15 years after that. It's a huge amount of change.

Jesse Jenkins, Princeton University (Milman, Chang, & Kamal, 2021, p. 2)

Europe will also need large investments in transmission capacity. In Denmark and northern Germany, for example, the growth in offshore wind generation has already outstripped the capacity of the transmission system to export northern wind generation to customer loads in the south. In the absence of grid capacity, wind generation must often be inefficiently curtailed, with one estimate of curtailment in 2018 in western Denmark equalling 5.7% of all electricity consumption in the region (Deign, 2020). Nevertheless, the EU plans large increases in renewable resources, with the share of electricity from renewables sources doubling from 32% in 2020 to 65% in 2030 (Amelang, Appunn, & Wettengel, 2020). This increase will be enabled not only by large investments in renewable generation, but also by large investment in grid interconnections to move electricity to customer load centers from wind farms in the Baltic and the solar farms of Spain (Amelang et al., 2020). This investment in transmission infrastructure, whether in Europe, the US, or any other region, will become ever more important to maintain system resilience with the growth of variable renewable resources.

The Constraint of the Business Model

Part 3 of this book describes constraints that the traditional utility business model places upon utilities in responding to a changing environment. This section will provide a brief preview of some of these constraints as they particularly apply to "System Decentralization" and "Transmission Infrastructure," two key elements of both resilience and decarbonization.

One of the basic premises of utility regulation is that a utility's profits are determined by the regulated return on the capital that a utility has invested in assets to serve its utility customers. This concept, known as "Cost of Service" regulation, which has been fundamental to the dominant utility business model since the 1930s, will be described in greater detail in Chapter 6. Under this framework, the profit earned by a utility is based upon the amount of capital invested in the electrical system, and not based on the services provided by the utility. However, this can pose an impediment to investment in system decentralization and transmission infrastructure.

Impediment to Investment in System Decentralization

Under traditional utility business models, distributed energy resources, such as rooftop solar generation or on-site batteries, are typically owned by the customer and not the utility. Since utilities have not invested their own capital in these assets, utilities have little economic incentive to support their wider deployment. Furthermore, by enabling customer self-generation of electricity, utilities will deliver less electricity to customers, resulting in rising rates for all customers. If left unchanged, traditional utility business models and regulatory structures will act as disincentives for greater decentralization of energy resources. These issues and their remedies will be reviewed in greater detail in Part 3.

Impediment to Investment in Transmission Infrastructure

The existing utility business model can act as an impediment to investment in enhanced transmission infrastructure. If utility business models are structured to generate profits based on the amount of capital invested, utilities will be incented to build resilience by investing in their own systems, through system hardening or development of redundant capacity, rather than by enhanced transmission connections to adjoining jurisdictions (Tsuchida & Gramlich, 2019).

Transmission system operators may also lack incentive to promote tools that would reduce transmission congestion. Although these tools have been shown to offer significant potential benefits, their actual deployment appears to be much less than expected, perhaps due to the misalignment of incentives of the regulatory model (Tsuchida & Gramlich, 2019). Operators of transmission systems are often compensated not upon improvements in operational efficiency, but rather on meeting minimum thresholds of efficiency. If minimum standards are met, the costs of congestion are simply passed through to consumers (Tsuchida & Gramlich, 2019). As a further disincentive, if technological tools are successful in reducing congestion, then new transmission investments could be reduced, with a corresponding reduction in returns to the owners of the transmission assets (Tsuchida & Gramlich, 2019).

Twenty years ago, large generation plants could take years or even decades to plan, approve, and construct, providing planners with long horizons to put in place needed transmission capacity. Today, however, the time frame for developing utility scale renewable generation is much shorter, with development periods sometimes of only a year or two. In addition, flow patterns on transmission systems are becoming more complex with increased penetration of renewable generation (Tsuchida & Gramlich, 2019). Although management of the transmission system has become more demanding, the utility business model has the potential to impede enhancements needed to improve system resilience. In response, regulators need to consider innovative policy options that alter the business model, such as new benefit sharing mechanisms that align incentives of the utility with desired outcomes. These regulatory responses will be reviewed in Chapter 10.

2 The Business Model

> They [business models] are, at heart, stories – stories that explain how enterprises work. A good business model answers Peter Drucker's age-old questions: Who is the customer? And what does the customer value? It also answers the fundamental questions every manager must ask: How do we make money in this business? What is the underlying economic logic that explains how we can deliver value to customers at an appropriate cost?
>
> Joan Magretta (2002, p. 4)

It is fascinating to look through past listings of members of the Dow Jones Average. Looking back a century into the composition of the index in the 1920s provides a reminder of names that reflect a different time and a different economy: The Central Leather Company, American Locomotive, and Remington Typewriter. However, much more recent history also provides examples of companies that have failed to adapt their business model to changes in their environment. In the thirty years since the 1990s, names like Bethlehem Steel, Sears Roebuck, and Kodak have disappeared from the Dow Jones Index through bankruptcy. Others, like Allied Signal and Texaco lost their independence through mergers or takeovers. Some have survived, like US Steel, General Electric, and F. W. Woolworth, but as much smaller enterprises and have been removed from the index due to diminished capitalization. These are all companies with decades of success, that as recently as thirty years ago were among the most powerful enterprises in the world's dominant economy. However, many of them failed to adjust to the world around them. Their business models no longer enabled them to compete against the best, and they have either disappeared or are greatly diminished.

The idea of the "business model" provides a useful framework to understand how a company adapts, or fails to adapt, to changes in its environment. The concept is used throughout this book to help understand how the electric utility industry has arrived at its current state, to understand the stresses it is currently under, and to anticipate innovations that utilities must consider in responding to a rapidly changing environment.

What Is a Business Model?

The business model is a description of the way a firm works to survive and excel. From earliest times, firms have implicitly operated according to some form of business model, even if the term was not used to describe how the business operated (Teece, 2010). The common usage of the term "business model" only came into use in the 1990s (Magretta, 2002) when it was sometimes used to explain the soaring valuation of a firm that was using the internet to conduct business in a new way. The promise of a new business model was, as the author Michael Lewis explained in 1999, "central to the Internet boom; it glorified all manner of half-baked plans" (Lewis, 1999, p. 256). As he further described, "The business model of most Internet

https://doi.org/10.1515/9783110714036-003

companies was to attract huge crowds of people to a web site, and then sell others the chance to advertise products to the crowds. It was still not clear that the model made sense" (Lewis, 1999, p. 257).

Despite the hype, the business model concept continued to be developed by firms who found it helpful in understanding their business and those of their competitors. One of the reasons the concept became more useful was the continued growth of personal computers and spreadsheet software, which enabled firms to efficiently analyze the cost and revenue structures of alternative business models (Magretta, 2002). With these computing tools, companies that could articulate their assumptions about their marketplace and the economics of their proposed business model could now quickly analyze multiple variations of a business model before the launch of a new enterprise, each with alternative supply chains, cost structures and pricing models (Magretta, 2002).

One of the strengths of the business model concept is that it is, by definition, a model. Like an architect's model of a building, the model of the business represents only those aspects required to be useful to the user (Baden-Fuller & Morgan, 2010). The model reduces the background noise of the business and forces managers using the concept to reduce the complexity of the business to key concepts. The act of developing the model and comparing alternatives adds clarity to the analysis of a firm's positioning and provides a structure that focuses management attention on key issues of the business (Baden-Fuller & Morgan, 2010).

The Elements of a Business Model

The business model is a conceptual framework that describes how a business is structured, and there are many different descriptions of the elements of a business model. Some writers describe over fifty different elements of a business model (Abdul Aziz, Fitzsimmons, & Douglas, 2008), some describe nine (Osterwalder & Pigneur, 2010), others use five (Matzler, Bailom, von den Eichen, & Kohler, 2013), many use four (Gassmann, Frankenberger, & Csik, 2014), while others have broken the business model into just two main parts (Magretta, 2002). Many of these alternative models offer similar concepts that primarily differ in granularity. Just as pizza can be cut into four slices or forty, so too can a business model. How finely a person wishes to dissect the business model depends largely on the requirements of the analysis.

Despite the varied descriptions of the business model concept, four main elements are commonly found across most descriptions (Baden-Fuller & Mangematin, 2013): an identification of the customer, a value offering to the customer, a value chain for delivery of products or services, and a means of value capture. In working with industry executives to develop the content for this book, these four elements were found to be useful to frame effective discussions about complex issues, while remaining detailed enough for insightful analysis. A four-part description of the business model based on a structure proposed by Baden-Fuller and Mangematin (2013) is used throughout this book (see Figure 2.1).

Customer Identification	Value Offering
Who is the firm targeting as a customer?	How does the firm deliver value to the customer, and what does it deliver?
Value Chain	Value Capture
How does the firm deliver the product or service?	How does the firm generate a profit?

Figure 2.1: The Elements of a Business Model.

A detailed description of each of these four business model components could merit books on their own, but are documented in summary as follows:

1. **Customer Identification:** Who is the firm targeting as an existing or new customer group? This may require market segmentation to better understand the market's distinct customer groups, and to better enable the firm to tailor value offerings to specific segments. It will identify potential new customers that the firm may choose to target and will include analysis of the existing customer base to identify markets that the firm may choose to exit. It may include a specification of whether the customer relationship is one-sided or two-sided, and if two-sided, a recognition that some of these customers may not be paying customers (Baden-Fuller & Mangematin, 2013). (Two-sided business models are discussed later in this chapter.)

2. **Value Offering:** How does the firm interact with customers to create a product or service that has value, and what service or product of value is delivered to the customer? Value offerings might be "bus" systems that are built for the economical delivery of a standardized product, or "taxi" systems, which are project-based and customizable (Baden-Fuller & Mangematin, 2013). The description of the value offering may overlap with the description of value capture, if favourable terms of payment or financing are included in the value offering to the customer. Some innovative firms do not just focus on following a me-too approach to delivering value, but instead have redefined what value is before the customer has requested it (Massa, Tucci, & Afuah, 2017). For example, Ikea made transportability of its unassembled furniture a key part of its value offering years before any other competitors (Massa, Tucci, & Afuah, 2017), while Nescafe revealed the convenience and individuality of Nespresso pods to consumers before they had demanded it (Matzler, Bailom, von den Eichen, & Kohler, 2013).

3. **Value Chain:** How does the firm fit in an ecosystem of suppliers to deliver the product or service to the customer? The value chain describes the mechanisms

used by the firm to deliver the product or service to the customer including, among other factors, decisions on vertical integration, suppliers, contracting, and networks (Baden-Fuller & Mangematin, 2013). A business model has also been described by some scholars as an "activity system" (Zott & Amit, 2010, p. 216) with a value chain in which partners, suppliers, customers, and others cooperate, often with a lead firm to coordinate activities. A fundamental part of the business model design is to determine in which elements the firm will participate directly, and in which elements it will coordinate with partners or suppliers to carry out key activities (Zott & Amit, 2010).

4. **Value Capture:** How does the firm generate a profit? For each of its customer segments, the firm needs to understand how it generates revenue that is consistent with other elements of the business model. For example, pricing may vary between market segments determined in Customer Identification. Pricing may also take into consideration the fixed and variable cost structure of the value chain (which, as discussed in later chapters in this book, has increasing relevance for electrical utilities). Pricing is often used by firms as a signal of quality in the value offering, as purveyors of luxury goods have discovered when distinguishing their products from competitors. The value capture element also considers such factors as customer financing, volume discounts, and whether prices are negotiated or firm for all parties (Baden-Fuller & Mangematin, 2013). As will be discussed in later chapters, a key element of value capture for a regulated monopoly service provider, such as an electric distribution utility, will include interaction with a regulatory body.

The Impact of a Dominant Business Model

Over time, the shared understanding of a business model can become dominant within a company and can even become embedded across an industry (Porac, Thomas, & Baden-Fuller, 1989). Developing a shared understanding of a dominant business model can be a powerful tool for a management team. This common mental framework helps management to filter irrelevant information to expedite decisions and to speed the implementation of those decisions (Tripsas & Gavetti, 2000). However, when a dominant business model is embedded in the mind set of a firm or an industry, it can also be an impediment to managerial creativity and can constrain the ability of management to innovate (Sosna, Trevinyo-Rodrgiuez, & Velamuri, 2010). Managers that become accustomed to working within a dominant business model can have difficulty visualizing new business models (Tripsas & Gavetti, 2000).

Take, for example, Polaroid, once a dominant firm in predigital photography between the 1960s and 1980s. Polaroid developed a successful business based on a "razor blade" business model, so named by the success of the inventor of the safety razor in the early 1900s, King Gillette. After initial disappointing sales in 1902, his first year of operations, with the sale of 51 razor blade handles and 168 blades, Gillette modified his business model (Matzler et al., 2013). Over the following three years he distributed razor blade handles at low cost or for free with packages of coffee, tea, or chewing gum, and made substantial profits with the sale of 134 million razor blades to fit those free razor blade handles (Gassmann et al., 2014). Thus, was born the "razor blade" business model. Polaroid utilized this business model with great success by charging a low price for its cameras but generating high margins from the sale of Polaroid film. Polaroid dominated the market for instant hard-copy photographs for decades but foresaw the coming threat to its business from digital photography. Polaroid invested heavily in the new digital technology in the 1990s, becoming a leader in the early technology surrounding digital photography (Tripsas & Gavetti, 2000). However, Polaroid's plans for commercialization of the new technology were developed through their existing "razor blade" business model, which was based upon an instant printed photograph as central to the value offering, and the main source of profit. This led Polaroid's technical research to focus on providing users with the ability to immediately view the image through a printed photograph, rather than a digital screen, as became the common practice (Tripsas & Gavetti, 2000). By the time that experience proved that an instant hard copy was not a necessary part of the value offering of a digital camera, the company's technical advantages had been overtaken by competitors, and Polaroid's position in the digital camera market was lost (Tripsas & Gavetti, 2000). The company's managers squandered their early favourable position in digital photography technology and their strong brand position in the marketplace by trying to fit the new technology into an old business model and failing to adapt their existing business model to the new digital technology. Polaroid filed for bankruptcy in 2001.

The example of Polaroid is intended to show that managers operating with a strong sense of a dominant business model can filter information and opportunities that fall outside of a dominant logic (Prahalad & Bettis, 1986). Polaroid's failure to capitalize on its technological capabilities in the 1990s was due to the rigid mental models that "discouraged search and development efforts that were not consistent with the traditional business model" (Tripsas & Gavetti, 2000, p. 1158). When top managers spend years developing knowledge of a core business, they will tend to also apply that perspective to new business environments (Prahalad & Bettis, 1986). Consequently, incumbent firms are more constrained than new entrants when confronted with opportunities or challenges that do not fit existing business models (Chesbrough & Rosenbloom, 2002).

The shared understanding of a business model can be dominant within a company and can also become shared and embedded across an industry (Porac et al., 1989). Like the old saying that "when you give a kid a hammer, everything looks like a nail," this shared perspective will tend to cause managers within an industry to approach new environments with old business models. This effect may be particularly found in mature industries, since research has shown that business models can be quite fluid in the early stages of a new industry but will become less fluid as the industry ages (Bohnsack, Pinkse, & Kolk, 2014). As discussed in later chapters, the prospect of management attempting to fit new technology into an old business model is an issue that the utility sector faces today, with incumbent business models reinforced both by management tendencies, and a decades-old regulatory structure.

Innovation of a Business Model

Companies need to be able to change and innovate their business model to survive and thrive (IBM Global CEO Study Team, 2006). IBM's interviews with 765 CEOs found that while business model innovation generally received less emphasis than product or operational innovation, those companies that displayed the greatest growth in operating margins placed twice as much emphasis on business model innovation than did underperformers (IBM Global CEO Study Team, 2006). Following decades of relative business model stability, electrical distribution utilities are experiencing increased pressure to innovate. Electric utilities are facing increased competition for the supply of electricity from nongrid sources (e.g., rooftop solar), and are facing increased pressure to rapidly decarbonize their own operations, and to support the decarbonization of other sectors of the economy, such as transportation. As is discussed in later chapters, each of these factors is creating an impetus to innovate the established utility business model.

The pressure to change and innovate business models applies to both new and incumbent firms. New, entrepreneurial firms are faced with uncertainty and use new business models to explore business opportunities (McGrath, 2010). Established firms, however, will tend to consider business model innovation only when changes in the external environment appear to threaten profitability (Sosna et al., 2010). Despite the significant benefits that may be available through business model innovation, businesses in mature industries can face significant impediments to business model experimentation. Whereas entrepreneurial firms can start with a "blank sheet of paper" in designing their business model, incumbent firms sometimes must deal with factors that impede change. Many of those factors that impede business model change have been studied by academics under the subject of "path dependence."

Path dependence has been widely discussed and used broadly in many areas of social sciences, including economics (Arthur, 1989; David, 1985); political science (Pierson, 2000) and business strategy. There are four bodies of research on path dependencies of particular relevance in explaining the difficulty that utilities sometimes have in changing their business model:
- Technological path dependence
- Embedded processes, relationships and values
- Shared strategic frames
- Financial returns

Technological Path Dependence

Studies of the implementation of new technologies have discovered that choices some-times made in the early implementation of a technology can lock in users to that tech-nology, even though the technology may later prove to be suboptimal. A famous early examination of this phenomenon studied the dominance of the QWERTY keyboard (David, 1985) which, despite its suboptimal functionality relative to competing config-urations, persists to this day in a position of dominance due to its large installed base of skilled users and QWERTY-configured devices. The configuration of the QWERTY keyboard reflects events that may have seemed insignificant at the time. For example, the top row of the QWERTY keyboard was probably designed to include all the letters required for a typewriter salesman to easily type the word "typewriter" when demon-strating to a prospective customer (David, 1985). This may have been helpful to an early typewriter sales department, but it did little to arrive at an optimum configura-tion for subsequent generations of users (David, 1985). Nevertheless, once a technol-ogy gains an early lead, it can gain efficiencies of scale, driving down costs, and leading to effective lock-out of competing technologies. Once an effective "lock-in" is achieved, even suboptimal technologies can be difficult to dislodge (Arthur, 1989). This difficulty in switching technologies can also lock in a company to a resultant business model. For example, as will be discussed later in Chapter 4, Thomas Edison's early commitment to technologies based on direct current[1] (DC) later made it difficult for his companies to shift to a business model with a value offering based on alternat-ing current (AC), even though the competing AC technology was superior in several key attributes.

Embedded Process, Relationships, and Values

Sometimes, the processes, relationships, and values that become embedded in an organization can, in themselves, become path dependencies, narrowing options available to an organization's management. Firestone, one of the most successful tire companies in the 1960s, responding in the 1970s to the introduction of new ra-dial tire based on new tire technology, was studied by Donald Sull as an example of an organization bound by its legacy processes and values (Sull, 1999a). From the experience gained from their operations in Europe, Firestone's managers had seen this new radial tire technology coming years in advance and, in fact, were very active in developing a response to the threat posed by the new technology (Sull, 1999a). How-ever, the scope of the company's competitive response was constrained by its prior suc-cess, values, and practices. The company was constrained by its commitment to protect

---

1 See the glossary for definitions of "Alternating Current" and "Direct Current."

existing employees by investing in ill-suited factories in communities where the company had deep roots, and by investing to maintain long-term, but ultimately low-profit, customer relationships (Sull, 1999a). Sull calls this condition "active inertia," when an organization follows established patterns of behavior, even when faced with significant changes in its technological environment (Sull, 1999b, p. 1).

This embedded behavior is also often reflected in a company's capital investment decisions. A firm that continuously invests in assets that extend and support its traditional operations, that is, "resource deepening," is also inherently deepening its own path dependence on those resources and embedded processes that they support (Karim & Mitchell, 2000, p. 1062). For example, the sunk costs invested in a utility's assets can form a financial dependence in a regulated environment, or the technical expertise developed in operating its assets may form an internal strength that can also become embedded in an organisation and become a barrier to change.

Shared Strategic Frames

Managers who hold a shared understanding of their firm's business model, can be thought of as a sharing a "mental model" of the industry, its competitive environment and their organization's position in it (Massa et al., 2017, p. 83). On one hand, the shared mental model can add value to a firm by guiding managers and focusing attention as they seek new opportunities for the firm (Chesbrough & Rosenbloom, 2002). On the other hand, as was discussed earlier in the case of Polaroid, it can cause managers to filter information and opportunities that fall outside this dominant logic and creates a path dependency that restricts a firms' business model options (Sydow, Schreyogg, & Koch, 2009). Consequently, the choice of business models available to a mature firm can be highly path dependent (McGrath, 2010), and incumbent firms are more constrained than new entrants when confronted with options that break prior business logic (Chesbrough & Rosenbloom, 2002).

Financial Returns

Attractive returns from an existing business model can force an incumbent to continue down a continuing path, even as its environment is changing. For example, Christensen (1997) found that incumbent firms are often able to recognize potential opportunities for changes to existing business models but are often bound from pursuing those changes by the short-term financial incentives of their existing business models. Often, the early profit margins available from exploiting a disruptive innovation through a new business model is much less than the margin available from their existing business. From the perspective of the incumbent, the new business requires new resources, offers unattractive pricing, and requires the development of relationships with customers that they do not currently have (McGrath, 2010). As capital and management attention are focused on those investments bearing the greatest returns, the business model innovation will be starved of attention or resources (Christensen, 1997). The financial basis of path dependence may also stem from avoidance of financial loss, with alternative paths made unattractive by financial factors such as switching costs, potential stranded costs, or loss of monopoly pricing positions (Sydow et al., 2009), all factors of particular relevance to electric utilities.

One example of financial returns as a source of path dependence is illustrated later in Chapter 5. The instance describes Thomas Edison, who had introduced the first, true electric utility business model in 1882, based on technology that utilized direct current (DC). When alternating current (AC) technology emerged that enabled transmission of electricity over greater distances and with lower line losses, Edison stubbornly stuck with DC technology because that was where his businesses had the strongest patent protection (Granovetter & McGuire, 1998). Since his patent position in AC technologies was relatively weak and offered fewer opportunities for profit (Granovetter & McGuire, 1998), Edison resisted altering his business model to incorporate the new technology. In the end, Edison's DC-based business model was buried by AC technology.

Multisided Business Models

Anyone perusing literature on business models will quickly become aware that there are dozens of generic business models (like the "razor blade" model discussed earlier) describing how different firms compete in their marketplace (Gassman et al., 2014). One variant that has not historically been widely discussed in the context of electrical utilities, but is recently being discussed more frequently, is that of the multisided business model. The term "multisided business model" is frequently used interchangeably with "platform business model," while a firm using the multisided business model is often referred to as a "platform" (Fehrer, Woratschek, & Brodie, 2018).

In any business model, a firm will target a particular group of customers. What is unique about a multisided business model is that the firm has at least two sets of customers, and the value created for one set of customers will be reliant upon transactions with the other (Baden-Fuller & Mangematin, 2013). In a predigital era, newspapers provided one such example of such a multisided model, as they would sell classified advertising to one set of customers, but also sold newspapers to another set of customers who would form the readership for the classifieds. In a digital world, many more businesses have leveraged the internet to bring together groups of users in a multisided business model. Many of these companies, like Google, Uber, Airbnb, and Alibaba, have become household names worth billions of dollars by building digital platforms that bring together two or more user groups. Google, for example, brings together advertisers with internet users searching for information. Uber and Airbnb bring together private owners of vehicles and accommodation with customer groups seeking transportation and accommodation, while Alibaba brings together millions of merchants with hundreds of millions of customers (Parker, Van Alstyne, & Choudary, 2016). Although Amazon was first launched with a single-sided business model to sell books on the internet, it later leveraged its distribution capabilities and computing infrastructure to create a multisided business model, enabling other businesses to utilize the Amazon platform to access a large, new customer base.

Using a multisided business model, a firm incurs costs to serve both sets of customers and, like a newspaper, may be able to generate revenues from both groups of users (Eisenmann, Parker, & Van Alstyne, 2006). However, firms in a multisided model often subsidize one set of customers at the expense of another. For example, those who are reading this paragraph in digital format may be using a version of a reader software developed by Adobe. Adobe has two main groups of customers in a two-sided business model: readers and content creators. Adobe makes its basic Reader software available free of charge to hundreds of millions of readers, which creates a valuable target community for its second group of users, content creators. Both customer groups receive value from the products and services provided from Adobe, but only the content creators, who pay for a fee for the use of the Adobe software, generate revenue for Adobe (Eisenmann et al., 2006).

As will be discussed in following chapters, for the last century electric utilities have used a single sided business model for the delivery of electricity to customers, as is natural for a product generated in a central station and distributed with a one-way flow of electricity to a customer. However, in recent years, the growth of distributed energy resources, such as rooftop solar, is causing some regulators and utilities to look more closely at multisided business models. Under this model, utilities would act as a platform to allow a customer with a rooftop solar panel to sell their surplus electricity to a neighboring consumer in a peer-to-peer transaction, with the transaction recorded in a blockchain-based trading system. This type of structure has been implemented in small-scale pilots to create a local energy marketplace in locations such as Brooklyn in the United States (Cardwell, 2017) and Freemantle, Australia (Sinclair, 2020). Although there are many regulatory and technical hurdles to clear before this is widely deployed, from the utility's perspective, the widespread deployment of this business model would mark a significant departure from its traditional business model. This will be discussed further in following chapters.

Part 2: **Utility Business Model Transformation –
A History**

To understand where the business model of the electric distribution utility is going, it is helpful to understand where it has been. Understanding the evolution of the industry can be a daunting task, given the confluence of change in technology, regulation, social change, and consumer expectations during the industry's existence. Nonetheless, the following chapters will do just that, with a description of the industry's past, present and emerging dominant business models.

– Chapter 3 will examine the period before the first public electric utility[1] went into operation in 1882. In this period, the emerging electricity industry was characterized by the sale of "private plants," small electrical plants and equipment used for the self-production of electricity on the customer's own premises, for the customer's own consumption. Although these companies were not operating as utilities, they laid important groundwork in developing technologies that would support later utility business models.

– Chapter 4 will examine the arrival of the utility business model, with the establishment of the inaugural North American public electric utility in 1882. The innovation of the new business model was built around an entirely new value offering: the delivery to the customer of lighting and electricity, rather than electrical equipment.

– Chapter 5 will examine the second dominant utility business model formed in the 1890s, with innovation focused on developments in the value chain. It utilized new technologies to centralize production, slash unit costs, and ushered in an era of competition that would eventually drive the first utility business model to extinction.

– Chapter 6 will examine the third dominant utility business model, largely established in the 1930s which, like the prior business model, continued to be characterized by centralized production and increasingly efficient distribution of electricity over an expanding grid. The innovation of this business model was built around value capture, with the development of monopoly franchises and regulation of the utility's rates and returns. This basic business model, with some variants, has dominated the industry for most of the last eight decades.

This section will primarily use the experience of the American marketplace to understand the evolution of the industry's dominant business model. The early history of the industry in the United States is dominated by private enterprise, although municipal ownership and public regulation came to be significant factors in the industry by the 1930s. Other countries developed along their own unique paths. For example, to use an observation of Joseph Schumpeter (McCaw, 2006) "unlike America, Germany

---

[1] In distinguishing between a supplier of electrical equipment, and an electric utility, this book follows the definition established by the US Department of Energy which defines a utility as "a legal entity that operates the distribution facilities for delivery of electric energy primarily for use by the public" (US Department of Energy, 2018f).

https://doi.org/10.1515/9783110714036-004

resorted to public enterprise, as well, at a comparatively early stage. Occasionally this led to conflicts but in general this form of 'municipal socialism' (and also action by provincial bodies and states) was welcomed by the manufacturing industry" (Schumpeter, 1939, p. 440). Britain also largely relied on the public sector to build out its electrical system, and by 1929 70% of its electrical system was publicly owned (McCraw, 2006). Britain's publicly owned "National Grid" was used as a model for other countries building out their system. Canada's electrical system followed its own path of development, with a mix of public and private enterprise, but generally closely followed US developments (Nelles, 2003). Each of these countries followed their own paths in determining the ownership of the electrical system. However, one of the characteristics of the "business model" as an analytic tool is that it is customer-centric, and not shareholder-centric. Accordingly, the reader will find that the discussion of the business model evolution in the American context will provide insight across systems of other countries, regardless of whether the ownership was primarily public or private. As the discussion will demonstrate, the evolution of the business model is not driven by characteristics of ownership, but rather by changes in other factors, such as technology, customer offerings, and value capture. The analysis of the industry's evolution will have broad application across all developed economies.

With Part 2 establishing the historical foundation of the industry, Part 3 will examine today's factors that are driving the development of a new dominant business model for the distribution of electricity.

3 Before the Electric Grid

The momentous laws of induction between currents and magnets were discovered by Michael Faraday in 1831–1832. Faraday was asked: 'What is the use of this discovery?' He answered: 'What is the use of a child – it grows to be a man.' Faraday's child has grown to be a man and is now the basis of all the modern applications of electricity.

Alfred North Whitehead, *An Introduction to Mathematics* (1911, p. 34)

In the Beginning

As the great business historian Arthur Chandler observed in his study of the formative years of modern capitalism, "Most histories have to begin before the beginning. This is particularly true for one that focuses on institutional innovation" (Chandler, 1993, p. 13). This is certainly also true in examining the innovations that mark the history of the development of the electric grid. The scientific study of electricity had long preceded the first installation of an electric grid in the 1880s. These investigations were carried out by countless curious inventors, researchers, and tinkerers, ranging from Benjamin Franklin with his fabled kite-experiments in the 1700s, to Michael Faraday and his investigation of electromagnetic fields in the 1800s (Munson, 2005). This scientific exploration enabled great electricity-based innovations of the early and mid-1800s such as the telegraph, the telephone, and the arc-light, and means of powering them such as the chemical battery or dynamo[1] (Munson, 2005). However, before electric utility grids became available, early innovators in the 1870s and 1880s who wanted a steady flow of electric current in a factory to run an electric motor, or to bring steady electric lighting into a wealthy home, would have had to rely on their own private power plant that they would purchase, install and operate for their sole use.

Although the idea of private ownership of the means of supplying one's electricity might seem novel today, it was not that unusual in the context of other services that businesses or homeowners used at that time. Owners of houses and businesses relied on their own furnaces for heat, and outside of major cities would have to arrange their own supply of fresh water and sanitation. Factories had a long experience with self-generated mechanical power using steam power or water-flow. Why wouldn't they use private power plants to self-generate electricity as well?

The installation of a private power plant could bring tremendous benefits to a business or homeowner. Imagine a mill or factory operating before the availability of electric power. Power from a central source such as a steam engine or water wheel would typically be distributed by a system of power shafts or belts throughout the factory, with illumination provided by daylight or natural gas jets (Smil, 2017). Distribution of

---

1 See the glossary for definition of "dynamo."

https://doi.org/10.1515/9783110714036-005

power to new parts of the mill or factory would be a major task requiring new belt-driven or shaft-driven machinery and worn or broken belts and drive machinery would ensure constant shutdowns for maintenance and repair (Smil, 2017). Any failure of the central source of mechanical power would bring the whole factory to a stop. Speed adjustments of individual components of the power system would be challenging without a shut down for reconfiguration. Replacement of these mechanisms with electric motors would allow a decentralized form of power that could be easily speed-adjusted in individual parts of the production system, while greatly reducing the risk from a single source of potential failure. It would eliminate the tangle of belts and pulleys running high above the factory floor, opening ceilings to natural light during the day, and with electric light when dark outside (Smil, 2017).

Or imagine homes or offices before the age of the incandescent electric lighting. Outside of major cities, they would be most often reliant for light on kerosene or candles, or in major cities they might be served by gas lighting. In the 1880s, each gas jet would produce a pale-yellow flame with the light equivalent of a 12-watt incandescent bulb (Hargadon & Douglas, 2001). In addition to the ever-present threat of asphyxiation (Wasik, 2006), the gas jets emitted a light soot that stained the surfaces of walls and furniture, and the threat of fire arose with alarming frequency (Hargadon & Douglas, 2001). For example, students using the main library at Harvard University were required to leave before it closed each evening at sundown, since the university's administration wouldn't install gas lighting in the library due to fire risk (Hargadon & Douglas, 2001). Arc lighting was also a technology that had been available for several decades, using a technology that arced electricity between conductors. Although this produced an intense light that was used in lighting streets and larger spaces, it was generally seen as too harsh a light for smaller spaces. In addition, the arcing process actually consumed the conductors, so the lamps had a relatively short life (Hargadon & Douglas, 2001).

Dominance of the Private Power Plant

A progressive business or fashionable home in the 1870s or 1880s requiring a steady supply of electricity would own a "private power plant," an installation that would typically include a coal powered, reciprocating steam engine, connected by belt drive to an electric dynamo or generator, plus distribution wiring and supporting control equipment. Alternatively, factories that had previously used water for mechanical power may have continued to use waterpower to drive the dynamo. The dominant business model of firms selling these private power plants before 1882 bore little resemblance to those of today's utilities, with many entrepreneurial companies competing in various areas of the market, including those founded by Charles Brush in America and Werner von Siemens in Europe. Thomas Edison's (see Figure 3.1) group of companies was one of the best known suppliers of private power plants in this era, and will be used in the rest of this chapter to illustrate a typical business model of firms operating in the industry at that time.

Figure 3.1: Thomas Edison, circa 1882, Age 35.*Source:* Photograph by unknown author, distributed on Wikimedia Commons under a CC-BY 4.0 license. https://en.wikipedia.org/wiki/War_of_the_currents#/media/File:Thomas_Edison_c1882.jpg (Accessed November 3, 2020).

Observed through the four-element business model framework examined in Chapter 2, Edison's business model for private power plants bore the following characteristics:

Customer Identification

Private generation plants were usually owned by businesses needing electric lighting or electric motors for their operations, but sometimes also by the wealthy who wanted to substitute gas lighting in their homes with much more fashionable and expensive electric lighting (Bakke, 2016). An early residential owner of one of Edison's private plants was the financier, JP Morgan, whose residence was the first in New York to be illuminated only by electricity (Jonnes, 2004). Although Morgan may have enjoyed the novelty of the electric lights, the noisy, dusty, and smoky coal-fired generator located in the basement and providing lighting to his New York mansion proved to be tremendously unpopular with those living nearby (Wasik, 2016). Morgan's neighbor, Mrs. James Brown, complained that the powerful steam engines and electrical generators next door caused her whole house to vibrate, and the fumes to tarnish her silver. Furthermore, the warmth of the generator housing attracted the neighborhood stray cats in the winter, and their great caterwauling led to further complaints (Jonnes, 2004). In the toniest of Manhattan neighborhoods, there must surely have been some interesting and uncomfortable conversations between Mr. Morgan and his neighbors.

Value Chain

The electric incandescent light bulb is probably Thomas Edison's most famous innovation (although it is recognized today that he was only one of several innovators leading to the workable incandescent light bulb) (Bakke, 2016). However, one of

Edison's most significant innovations was in developing the capability to manufacture and deliver a complete electric system, including generation equipment, distribution equipment, and the light bulbs themselves (Munson, 2005). These components were first developed in Edison's Menlo Park labs, and then manufactured by a series of Edison controlled companies. For example, Edison Electric Lamp Company manufactured light bulbs, while Edison Machine Works built electric generators and motors (Wasik, 2006). This enabled Edison to deliver an entire electric system to a customer, often by rail, complete with the installation crew to get the system up and running. Edison was not the only developer of the private power plant, but he was the first to offer a fully integrated solution (Bakke, 2016).

Value Offering

Edison offered his "private central stations" as complete systems for factories, businesses, and homes of the wealthy. These systems would be customized, dependent upon their required use. Wealthy owners of mansions might desire electricity of a certain voltage and frequency for lighting, while owners of electric trolley companies might require different standards to power streetcars (Bakke, 2016).

Revenue Model

Edison's revenue model was largely transactional, reliant upon the sale and installation of the complete electrical system, with secondary revenue generated by the sale of replacement and expansion components. Since so many of the components were developed in Edison's labs for the express purpose of fitting this system, many of the components were proprietary, ensuring that replacement components came from Edison (Bakke, 2016). If something broke, the customer needed parts from Edison to fix it, and unless the customer had trained their own maintenance workforce, they also needed Edison's people (Munson, 2005). In addition, new market entrants attempting to integrate with Edison's system were required to pay royalties for the use of Edison's patents (Munson, 2005).

In summary, the pre-utility business model of the electric industry in the period prior to 1882 was dependent upon the provision of electrical equipment for the customer's own generation of electricity, not upon the provision of electricity (see Figure 3.2). The first utility for the distribution of electricity had not been introduced. Customers purchased these private plants to generate electricity for their own consumption, and the customer relationship and revenue models were primarily transactional. However, since Edison's early light bulbs had a relatively short life and carried a replacement cost of about $23 in today's dollars (Bakke, 2016), one supposes that the unreliability of

Pre-Utility Business Model
Co-location of Production & Consumption

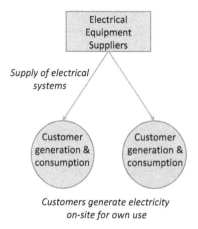

Supply of electrical systems

Customers generate electricity on-site for own use

Figure 3.2: Pre-Utility Business Model.

the early electric systems must have required a close and frequent relationship between supplier and customer.

Stretching the Private Plant Business Model

By mid-1878, Thomas Edison had developed one of the first versions of the incandescent light bulb that was both sufficiently bright and robust to be promising for commercial development. This new lighting technology naturally fit the private power plant business model, serving single customers with lighting energized by self-generation. In fact, this business model was very successful and remained successful for several decades. Even though Edison would launch a new, competing business model based on the "central station grid" in 1882 (and which we discuss in the next chapter), several of the shareholders in Edison's companies encouraged the continued development of private plants. In fact, this product would remain the Edison companies' largest revenue and profit source for many years (Jenkins, 1982), with the sale of over twelve hundred private power plants in the five-year period between 1882 and 1887 (Bakke, 2016).

As successful as the private plant business model may have been, Edison appears to have soon developed a vision of using the technology in a totally new business model: the central station grid. As noted in the 1910 biography of Edison, "From first to last Edison has been an exponent and advocate of the central-station idea of distribution now so familiar to the public mind, but still very far from being carried out to its logical conclusion" (Dyer & Martin, 1910, p. 345). Edison's vision was of a business model that would be based upon a system of generation in a central location, but with electricity distributed by electrical conductor to multiple sites and to multiple customers, even very small customers who might not have been able to purchase a full central station plant.

Achieving this business model innovation would require some extensive preparatory work. In addition to redefining the customers for his business and the value offering it would deliver, Edison would also need to develop technologies and procedures to support a new value chain capable of servicing multiple customers from a single plant, and a new way of capturing revenue from customers. He would also need to convince investors and potential customers of the viability of this vision. However, Edison had the benefits of a successful business in private power plants that he would leverage to build many of the new capabilities required for the new business model. Over the next several years, he would use many of the installations of his private plants to not only demonstrate to potential investors and customers the viability of distribution of electricity, but also to work out the supporting aspects of new technology required to support the new business model.

The first such demonstration was to be in 1879 at his own research facilities in Menlo Park, New Jersey, where, from an electric dynamo in his laboratory, he demonstrated to investors the illumination of fifty-three Edison incandescent light bulbs located in adjacent homes and on a few imaginary streets (Allerhand, 2019). Although investors apparently remained sceptical, one of those attending the demonstration was the businessman Henry Villard, the president of the Oregon Railroad and Navigation Company, who could see the potential for the new technology in one of his company's new steamships, the *SS Columbia* (see Figure 3.3) (Skjong, Rodskar, Molinas Cabrera, Johansen, & Cunningham, 2015). When the ship launched in spring of 1880, it had been fitted with the first installation in a commercial venue of Edison's incandescent electric lighting system (Allerhand, 2019).

Figure 3.3: *SS Columbia* (1880).*Source:* Photograph by unknown author, distributed on Wikimedia Commons under a CC-BY 4.0 license. https://en.wikipedia.org/wiki/SS_Columbia_(1880)# (Accessed November 13, 2020).

Although the installation of shipboard electrical lighting was probably not the deployment of his new lighting system that Edison had imagined, it did force his team to work out many of the technical aspects of electricity distribution that would later be used to support Edison's new central station business model. The SS *Columbia* was outfitted with a 120 bulb Edison lighting system, distributed via several circuits from four 6 kW dynamos (see Figure 3.4), belt-powered from the ship's steam engine (Skjong et al., 2019). This installation forced the development of new operational and technical solutions to manage the distribution of electricity from a central location across multiple circuits. For example, instrumentation to monitor system voltage had not been included by Edison in the delivered system, so operators would be required to adjust voltage by watching the brightness of light bulbs in the engine room. A further improvisation was the use of small lead wires that secured the circuits to also function as fuses (Skjong et al., 2019). In addition, the operation of the light switches could not be entrusted to passengers, so light switches were located in locked wooden boxes. If the lights were to be turned on or off, a passenger would call a ship's steward to unlock the box and operate the switch (Skjong et al., 2019). Fortunately, today we do not have to contact our electrical utility today to turn off the lights.

Figure 3.4: Edison's Belt Driven Dynamo, 1882.*Source:* Photograph by unknown author, distributed on Wikimedia Commons under a CC-BY 4.0 license. https://commons.wikimedia.org/wiki/File:Edi son_dynamo_1882.png (Accessed May 4, 2021).

The installation of the lighting system at Menlo Park and on the *SS Columbia* were foundational steps to the development of a new business model for the distribution of electricity. An important next step would be the first commercial installation of an Edison lighting system on land, at a New York firm of color printers and lithographers, Hinds, Ketcham & Company, that had previously been only able to work during daylight hours, due to the difficulty of matching colors of finished products under other forms of artificial light (Dyer & Martin, 1910). The new installation was an apparent success, with the customer attesting to the value of the lighting, "It is the best substitute for daylight we have ever known, and almost as cheap" (Dyer & Martin, 1910, p. 182). This installation on Water Street in New York followed Edison's existing business model for private power plants, with a single customer using the lighting delivered by a dynamo in the basement (Sulzberger, 2010). However, the location of the site, close to New York's financial and newspaper district, would act as a showcase site to attract new customers. As his biographers noted, Edison later explained, "We were very anxious to get into a printing establishment. I had caused a printer's composing case to be set up with the idea that if we could get editors and publishers in to see it, we should show them the advantages of the electric light" (Dyer & Martin, 1910, p. 182).

Edison would sometimes modify this private plant business model to address a business opportunity presented by an installation. For example, some of Edison's private plant installations started to provide surplus electrical power to adjacent neighboring businesses. One early example was that of a Boston printing company that arranged to sell surplus evening electricity from its newly installed Edison private power plant to the neighboring Bijou theatre. The story is told of Edison visiting the theatre for its inaugural performance in early 1882, the first under electrical lighting in the United States, and noticing the lighting start to dim during the performance (Stross, 2008). Edison and a colleague went to the printing plant next door to investigate and found the responsible boilerman repairing a steam leak. When other members of Edison's group followed, they found Edison and his colleague with their swallowtail jackets off, shoveling coal into the steam power plant. With vigorous steam generation restored, they returned to the theatre and to a much brighter production (Stross, 2008). As he was observing the dimming of the lighting in the Bijou Theatre, or later shoveling coal next door, Edison surely must have seen the potential of the new central station business model that he was already developing in New York (and discussed in the next chapter) that would supply electricity to multiple customers from a single, well-managed generation source, and remove from customers the requirement to fix steam leaks or shovel coal to obtain electricity.

New technologies and processes would be needed to enable Edison to service more than one customer at some distance from a single plant. The technical feasibility of the idea was tested in London with a lighting system on the Holburn Viaduct. This location had advantages, since stringing cables under the viaduct avoided the expensive and time-consuming requirement to dig up city streets, and it provided promotional opportunities since it was within minutes of the newspaper offices of Fleet Street (Electricity Council, 1987). The project, described as a "commercial experiment" (Electricity Council, 1987, p. 18) did operate successfully for about two years. As Edison's biographers described it, the plant at Holborn was "not permanent," but it did " . . . put into practice many of the ideas now standard in the art, and secured much useful data for the work in New York" (Dyer & Martin, 1910, p. 204).

As Edison was developing the technology to support the central station plant, his business continued to be faced with healthy demand for the private plant. As Edison's biographers noted:

> In this instance, demands for isolated plants for lighting factories, mills, mines, hotels, etc., began to pour in, and something had to be done with them. This was a class of plant which the inquirers desired to purchase outright and operate themselves, usually because of remoteness from any possible source of general supply of current. It had not been Edison's intention to cater to this class of customer until his broad central-station plan had been worked out, and he has always discouraged the isolated plant within the limits of urban circuits; but this demand was so insistent it could not be denied, and it was deemed desirable to comply with it at once, especially as it was seen that the steady call for supplies and renewals would benefit the new Edison manufacturing plants.   (Dyer & Martin, 1910, p. 182)

To meet this demand, Edison would form a separate company to pursue the continued sale of private isolated plants, even as he was pursuing the continued development of the central station grid business model (Dyer & Martin, 1910). The business model for the supply of private power plants (see Table 3.1) would remain viable for several decades, with private plants still the most common source of electricity in the United States as late as 1915 (Granovetter & McGuire, 1998). In fact, the business model has continued to exist in some form until today, particularly for customers in remote areas without access to the grid.

The implementation of many Edison private plants stretched the capabilities of the technology that was then available. The installations at Menlo Park, on the *SS Columbia*, on the Holborn Viaduct, the Bijou Theatre, and so many others required new technologies and processes. Some installations targeted new customer groups, like the private plant at Hinds, Ketcham & Company, which would demonstrate the value of electric lighting to a potential printing and publishing customer base. These plus many other private station installations supported the development of capabilities that Edison would use to launch his new venture on Pearl Street in New York. Edison was preparing to launch his new business model: the central station grid.

Table 3.1: Summary of Pre-Utility Business Model – Pre 1882.

Business Model Element		Description
Key Characteristics	–	Sale of equipment for self-production and self-consumption of electricity.
Customer Identification	–	Commercial users of electricity, such as trolly companies, arc lighting companies, or factories large enough to afford a dedicated electrical power plant. Also, individuals wealthy enough to install residential electric lighting.
Value Offering	–	Enable customers to self-generate electricity for own consumption, generally for lighting or electric motors.
Value Chain	–	Companies manufacture electrical equipment for self-generation of electricity by private customers, for self-consumption.
Value Capture	–	Transactional, based on sale of equipment and replacement components, plus installation, maintenance, and repair services.

Unique Conditions of the Electric Utility Industry

Technology had progressed by the early 1880s to the point where the building blocks for the first utility were in place. The following chapters will describe the utility business models that evolved from these early foundations. However, there are factors unique to electricity markets that should be considered in understanding these subsequent business models.

First, unlike almost any other product, the supply chain for electricity on a utility grid does not include an inventory component. As power system engineers learn early in their education, although coal can be stored as a fuel stockpile or water kept behind a dam for future generation of electricity, supply and demand of electricity must always be kept in balance once generated and in the electrical system (Muir & Lopatto, 2004). On the supply side, electricity is delivered from the point of generation to the point of consumption at close to the speed of light (297,000 kilometers per second) and is not yet economically stored at grid level with today's technologies in large quantities. On the demand side, consumption of electricity is constantly changing as millions of users alter their usage of everything from light bulbs to massive electrical motors. Response to over or under-supply conditions must be rapid, since today's electrical grids typically have only seconds of cushion built into the system. In North America, about 140 regional authorities in the United States and Canada maintain this incredibly complex balance between supply and demand within a narrow band of tolerance (National Research Council, 2010). Small variations are normal, as supply constantly adjusts to meet fluctuations in demand (Muir & Lopatto, 2004). However, an imbalance of supply and demand that is too far outside its prescribed level, can cause the frequency of the AC electrical system (which is 60 hertz [Hz] in North American systems) to increase when supply exceeds demand, or decrease when supply fails to meet demand. Large imbalances can cause rotational speeds of generators to fluctuate, causing damage or destruction of equipment on the system (National Research Council, 2010). A period of low system frequency can cause brownouts and automatic "load shedding," cutting blocks of customers from the system to avoid total system collapse (Muir & Lopatto, 2004). The cost of large imbalances leading to blackouts can be massive in today's economy. On 14 August 2003, a blackout impacted an area with estimated 50 million people in the American Midwest, the American Northeast, and Ontario, Canada. The economic impact of this disruption to the United States was estimated at between $4 billion and $10 billion, while manufacturing shipments in Ontario were down $2.3 billion during the month of the disruption, and Canadian gross domestic product declined by 0.7% in that month (Muir & Lopatto, 2004).

A second unique factor affecting the electricity market is that there is virtually no free market in electricity. Governments get involved to affect costs and prices, and even so-called deregulated markets receive oversight from regulatory bodies. Even Thomas Edison's very first utility, which we discuss in the next chapter, was required to pay the city of New York a fee for each mile of distribution wiring installed, and royalty based on sales (Hargadon & Douglas, 2001). Furthermore, utilities are often used by policy makers as tools to pursue social issues, such as economic development or income inequality. And electricity generation itself often creates market externalities from factors such as GHG production from fossil fuels, disposal of spent nuclear fuel, or end-of-life disposal of wind turbines and photovoltaic cells. Increasingly, governments are seeking to address these market inefficiencies by subsidizing alternatives (e.g., rooftop solar subsidies) or increasing the cost of the delivered product (e.g., carbon taxes).

A Primer on the Structure of the Electric Utility Industry

The next chapters discuss the evolution of the utility business model and its value chain components. Not all changes to the business model involve the "hardware" of the electrical system. Indeed, we will find that many of the most significant innovations have arisen in the business model's value offering, value capture and customer identification, areas with little or no impact on the hardware of the value chain. However, to prepare readers who have not previously had exposure to the hardware components of an electric utility, the purpose of this section is to provide a brief description of the industry's basic value chain, to enable a better understanding of the description of the industry business model. Those readers who already have a knowledge of the industry may wish to move ahead to the next chapter.

A traditional description of the basic electric system consists of three principal interconnected elements: a generation source, a transmission system, and a distribution system (see Figure 3.5). In addition, the function of "retailing" may be carried out by a separate entity in a deregulated jurisdiction. These are described as follows:

Generation: In a typical electrical system, electricity is centrally generated at voltages ranging from 10 kV to 25 kV. Central generation sources can employ a range of fuel sources, including nuclear, natural gas, coal, hydro power, wind, geothermal, and solar. The ownership of the generation assets will vary based on the regulatory and market structure in the jurisdiction. In the case of a fully integrated utility, the generation will be owned by the same utility that distributes it to the end customer. Alternatively, the generation source may be owned by an independent power producer (IPP), and in the case of very large consumer of electricity, the customer itself may own the generation asset (Muir & Lopatto, 2004).

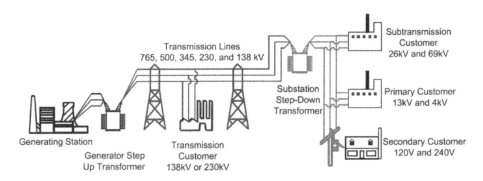

Figure 3.5: Simplified Structure of the Electric System.
Source: US–Canada Power System Outage Task Force, 2004.

Transmission: Once generated, electricity is "stepped up" through a transformer to higher voltage for transportation to centers of consumption, at distances ranging from a few kilometers to hundreds, and recently even thousands of kilometers. Typical transmission lines operate at 765, 500, 345, 230, and 138 kV, a much higher voltage than distribution lines to reduce the losses of electricity from conductor heat. Transmission of electricity has historically utilized alternating current (AC[2]) systems,

---

2 See the glossary for definitions of AC and DC systems.

although technological advancements in recent years have resulted in the increasing use of direct current (DC) systems for transmission over longer distances. Ownership of the transmission system may rest with the integrated utility in some jurisdictions, or it may also be owned by an independent operator. Some electricity may be drawn directly from the transmission system by larger industrial and commercial customers such as factories, processing plants, or large commercial facilities at intermediate power levels typically between 138 kV and 230 kV, with the balance delivered to customers through the distribution system (Muir & Lopatto, 2004).

Distribution: Electricity transported over the transmission system at high voltage will eventually reach a substation located in a demand center, where its voltage will be "stepped-down" (i.e., reduced) to a voltage that is useable by customers on the distribution network. Residential customers on the distribution network will typically take electrical service at a relatively low voltage of 120 or 240 volts in North America or 220 volts in Europe, while larger commercial or industrial facilities may connect at higher voltage levels (typically 12 kV to 115 kV) (Muir & Lopatto, 2004).

Retailing: In a fully integrated, traditionally regulated electrical utility, the function of "retailing" may not exist as a separate unit in the organization chart but may instead be found embedded in the functions of the utility's other business units, such as customer service or distribution. However, in jurisdictions that have undergone deregulation (discussed further in Chapter 6), the customer-facing activities of marketing, billing, and customer contact, together with energy procurement, are sometimes moved to third-party "retail service providers" who will buy electricity from generators and sell to end-use customers in a competitive marketplace.

This illustration is an overly simplified representation of the many different electrical systems that make up "the grid." One of the complicating features of the grid that has emerged in importance in the past decade, and that receives particular attention in Part 3 of this book, is the growth of distributed energy available from devices such as rooftop photovoltaics, wind turbine generators, electric vehicles, or connected battery systems. These distributed energy resources connect to the distribution system rather than the transmission system, and each has an output capacity that is much less than a typical central generation unit. However small these energy sources may be, their proliferation has greatly increased their importance in recent years in many jurisdictions, and the move to ramp up renewable energy in the coming years will increase this proliferation and importance even further. For example, in order to meet its objective of 100% renewable energy, the state of Hawaii plans by 2045 to have placed distributed photovoltaics (PV) on one quarter of the roofs of commercial customers, and on all residential customers (Allan & Pang, 2020). This decentralization, discussed in later chapters, will represent a significant transformation from the traditional structure of the electrical system described here, and a significant management challenge to utilities, policy makers and regulators.

4 The First Utility Model – The Central Station Grid

This truly significant achievement at Pearl Street was not immediately recognized for the milestone it was – the first truly complete and integrated electric utility system.

Robert Noberni, *IEEE Power Engineering Review* (1982, p. 2)

The First Central Station Grid

While many entrepreneurs in the 1880s were striving to use advances in electrical equipment to develop commercially successful ventures, it was Thomas Edison who combined this technology with farsighted business vision to create a new business model, and the first viable version of today's electricity distribution systems. As early as 16 September 1878, in an interview with the *New York Sun* to describe how his laboratory had just solved the problem of incandescent lighting, Edison was already laying out a remarkable vision for the supply of electricity from a central plant:

> Again, the same wire that brings the light to you," Mr. Edison continued, "will also bring power and heat. With the power you can run an elevator, a sewing machine or any other mechanical contrivance that requires a motor, and by means of the heat you may cook your food. To utilize the heat, it will only be necessary to have the ovens or stoves properly arranged for its reception. This can be done at trifling cost. The dynamo-electric machine, called a telemachon, and which has already been described in *The Sun,* may be run by water or steam power at a distance. (Rutgers, 2020a)

Although the term "telemachon" seems to have been lost in history, Edison would spend the next four years applying lessons learned in distributing electricity from private plants, such as onboard the *SS Columbia* or his Menlo Park facility, to fulfill that vision. The result in 1882 was the opening of New York's Pearl Street station, generally referred to as the first public electric grid in North America, and the first central station delivering incandescent lighting to multiple customers (Noberni, 1982).

The Edison Electric Illuminating Company (a direct-line ancestor of today's Consolidated Edison, or "ConEd," still serving the City of New York) was formed in December 1880 to undertake the project, and received a franchise from New York in April of 1881 to lay electrical distribution cables under the streets of the city. Months of planning, financing, and construction followed, and on 4 September 1882 the Pearl Street plant commenced operations (Sulzberger, 2010). The plant only occupied the footprint of two standard city lots, at 50 feet by 100 feet, but it contained four coal-fired Babcock & Wilcox boilers in the basement, powering six reciprocating steam engines and Edison-built "Jumbo" dynamo units on the ground floor. The Jumbo dynamos (see Figure 4.1), named after the famous circus elephant owned

https://doi.org/10.1515/9783110714036-006

by P. T. Barnum, had been purpose-built for Pearl Street based on an earlier design by Werner von Siemens (Sulzberger, 2010). These fed electricity into an innovative underground distribution system with 14 miles (about 22 km) of conductor buried under the adjacent streets of Manhattan and, to support and manage it all, Edison developed new and purpose-built controls and equipment (Noberni, 1982). The grid serviced by this station when it opened was a little more than one- sixth of a square mile in Manhattan's financial district and grew at full expansion to service a territory of about 1 square mile by 1884 (Stana & Apse-Apsitis, 2015).

Figure 4.1: Edison's "Jumbo" Dynamo. *Source:* Image courtesy of Museum of Innovation and Science, Schenectady, New York

This first central grid at Pearl Street represented a remarkable innovation of the prior "private stations" business model. Edison's decision to change the focus of his value offering from electrical equipment to electricity and light would require innovative redefinition in all other major components of the business model. Let us look at the innovation in those components.

Innovation in Value Offering

First, and perhaps most importantly, Edison changed the value offering presented to customers who desired lighting and electricity. Before 1882, Edison might have sold these customers, such as JP Morgan, electrical equipment for self-supply of electricity. Now, however, Edison's vision was that electrical equipment should be built and sold to central generation stations, and not to the end consumers of electricity. Instead, the value offering was totally changed for customers who desired lighting: the product sold from these generating stations to businesses and homeowners should be electricity and light (Granovetter & McGuire, 1998). This was a value offering that had not previously existed.

For customers accustomed to kerosene and gas lighting, one might imagine immediate pent-up demand and an easy sale for such a product as electric incandescent lighting. However, communicating the value of this new offering to a customer who has never used it, was not without its challenges. As Edison's 1910 biography noted:

> Nothing is more difficult in the world than to get a good many hundreds of thousands or millions of people to do something they have never done before. A very real difficulty in the introduction of his lamp and lighting system by Edison lay in the absolute ignorance of the public at large, not only as to its merits, but as to the very appearance of the light. Some few thousand people had gone out to Menlo Park, and had there seen the lamps in operation at the laboratory or on the hillsides, but they were an insignificant proportion of the inhabitants of the United States. (Dyer & Martin, 1910, p. 330)

Even if potential customers were aware that electricity delivered from a central station grid was available in their communities, actual experience with the product and its value offering was necessary to drive demand. As Edison's biographers noted:

> Of course, a great many accounts [of electricity delivered on Edison's electric grid] were written and read, but while genuine interest was aroused it was necessarily apathetic. A newspaper description or a magazine article may be admirably complete in itself, with illustrations, but until some personal experience is had of the thing described it does not convey a perfect mental picture, nor can it always make the desire active and insistent. Generally, people wait to have the new thing brought to them. (Dyer & Martin, 1910, p. 330)

Fortunately for Edison, the value of incandescent light was eventually proven, with superior cost and functional benefits when compared to gas or kerosine lighting. With the cost of infrastructure shared across many customers, electricity from the central station grid would prove to be typically cheaper than electricity from private power plants. In addition, electric incandescent lighting was cleaner, safer, and brighter than kerosene or gas lighting alternatives (Nelles, 2014). In 1879, Edison had been quoted in the *New York Herald* that "we will make electricity so cheap that only the rich will burn candles" (Rutgers, 2020b). Although this boast did not quite come to pass, he certainly did eventually demonstrate savings to some customers. For example, when The Herald itself connected to the Pearl Street grid in the spring of 1883, Edison's managers estimated that the newspaper's annual gas lighting costs would be cut by 60% (Wasik, 2006).

Innovation in Customer Identification

The business model for Edison's existing private station plants targeted customers across the country and internationally who wished to self-generate electricity for their factory or mansion. However, under Edison's new business model, the potential customers of the Pearl Street station (see Figure 4.2) could only be those within the distance that a low-voltage electrical current could be transported on an electrical feeder,[1] which was about 1.5 miles from the station (about 2.5 km) (Stana & Apse-Apsitis, 2015). As discussed later in this chapter, this was a technical constraint then faced by Edison or anyone else trying to send low-voltage currents through an electrical feeder, although new technology extended this reach a few years later to about 6 miles (about 10 km) (Millard, 1992).

EXTERIOR OF THE PEARL STREET STATION
Edison's first generating plant.

Figure 4.2: Pearl Street Station.
Source: Image courtesy of Museum of Innovation and Science, Schenectady, New York

The Pearl Street location was selected by Edison for several reasons, not least of which was the type of customers that it could access. First, the station was located in a densely populated area of Manhattan, with a mix of potential residential and commercial customers. Second, the area included what was, and still is, the financial center of the United States, with potential customers such as the offices of JP Morgan. Finally, the area was home to many of New York's newspapers, including *The New York Times,* and favorable publicity of the new electrical service would be necessary to create interest among both new customers and financial backers (Sulzberger, 2010).

---

1 A "feeder" generally refers to the electrical cables that connect consumers of electricity to the distribution substation.

On 4 September 1882, the head electrician at Pearl Street threw the master switch, and for the first time the plant connected to 85 customers with 400 light bulbs (Noberni, 1982), including 106 light bulbs at the offices of JP Morgan and 52 light bulbs at *The New York Times* (Wasik, 2006). The venture quickly grew its customer base, and within two months of opening, the Pearl Street station had 203 customers, and 513 within another year (Wasik, 2006).

Innovation in Value Chain

Edison had leveraged his years of experience in installing private power plants in locations like the *SS Columbia* and the Holburn Viaduct to develop new technologies and procedures that would be used in Pearl Street. Nevertheless, despite this experience, the extent of development of new technologies and procedures still required was remarkable. As Edison's 1910 biography noted:

> Edison had resolved from the very first that the initial central station embodying his various ideas should be installed in New York City, where he could superintend the installation personally, and then watch the operation. Plans to that end were now rapidly maturing; but there would be needed among many other things – every one of them new and novel – dynamos, switchboards, regulators, pressure and current indicators, fixtures in great variety, incandescent lamps, meters, sockets, small switches, underground conductors, junction-boxes, service-boxes, manhole-boxes, connectors, and even specially made wire. Now, not one of these miscellaneous things was in existence; not an outsider was sufficiently informed about such devices to make them on order, except perhaps the special wire.
>
> (Dyer & Martin, 1910, p. 331)

A key driver for the number of required new technologies was the extent of integration that the new enterprise would undertake. To deliver lighting to a new customer, Edison would leave the construction of boilers and steam engines to others, but his companies would build and install the generating equipment, build the control infrastructure, install the distribution feeders under city streets, wire the buildings, install the light bulbs that they had manufactured, establish operating procedures to run everything, and even show up later to change customers' burnt-out light bulbs. Throughout all phases of Pearl Street's design and construction, Edison provided the integrating vision and acted as the project's chief engineer (Sulzberger, 2010). It is hard to find an example of another innovation that has required such a degree of integrated value chain innovation before the first customer has been serviced or the first dollar or revenue earned.

Another reason that so much innovation was required throughout the value chain was that rather than focusing on building separate components, Edison's business model focused on delivering a complete system (Hughes, 1993). This approach could be seen in other inventions pursued by Edison, where he looked at innovation through the entire value chain to the customer (Hughes, 1993).

Take for example, his lesser-known inventions with Portland cement, where he held forty-nine patents throughout the value chain for the building material (Rutgers University, 2020c), from processing equipment, paint for waterproofing concrete, to molds for single-pour concrete construction of single-family homes[2] (Heun & Moss, 1987). This focus on inventing systems rather than components was also reflected in the development of Pearl Street Station electrical grid. Edison did develop a commercially useable incandescent light bulb, but he had no more claim to the invention of the incandescent bulb than other inventors such as Joseph Swan (Hughes, 1993). He also was not the sole inventor of the electrical generators nor much of the distribution systems that would energize those bulbs (Hargadon & Douglas, 2001). However, Edison had an ability to think in terms of systems, so combined his light bulb with generators, feeders, and distribution system to deliver lighting to the end consumer, to much greater success than ever achieved by inventors of individual components (Hughes, 1993).

Innovation in Value Capture

Finally, instead of a value capture approach that focused on revenue from the sale of electrical equipment, Edison would focus on the generation of revenue using an early form of consumption-based, volumetric pricing. With accurate electrical metering technologies not yet developed, Edison nevertheless charged Pearl Street customers for their consumption, by charging "by the light bulb" (Wasik, 2006). Although this was an imperfect method of matching revenues and costs and was replaced in following years by more accurate methods, it did provide customers and supplier with an initial means of measuring service and agreeing a corresponding unit price.

A Business Model Constrained by Distance

As we learn in high school physics, there are two basic means of distributing electrical power: alternating current (AC) and direct current (DC). DC electricity, from a source like a household battery, runs in a single direction, while AC electricity,

---

2 Those fortunate enough to have attended a baseball game in the original New York Yankee Stadium before its demolition in 2008, sat in a structure built in 1922 with Edison Portland cement. When the stadium was renovated in 1973, its concrete walls were so durable that they were left untouched (Heun & Moss, 1987).

reverses direction many cycles per second, measured in Hz.[3] In the years leading up to the opening of Pearl Street, Edison had focused much of his research on DC-based technologies.[4] Due to this research focus, Edison always had his strongest patent protection in DC generation and distribution, while his patent position in AC technologies, on the other hand, was relatively weak and offered fewer opportunities for profit (Granovetter & McGuire, 1998). As would be reasonably expected given his relative advantages, Edison used DC-based technology for the development of his central station grid.

DC power also held certain operational advantages in the early 1880s. For one thing, batteries often formed part of the installed electrical system, and could be directly charged and discharged using DC current without the need for additional equipment to convert the current to or from AC (Roguin, 2004). In addition, although incandescent lighting could operate under either AC or DC, a workable AC-based electric motor had not yet been developed in the early 1880s and would not be until later in the decade (Roguin, 2004). Rudimentary DC motors, on the other hand, had been around for some time. As early as 1839, Moritz Hermann von Jacobi, a Prussian engineer and inventor, had run an electric motorboat for 1.8 hours using a battery-powered 1 kW DC electric motor (Molinas & Monti, 2017). Because of this early development of DC motors, environments requiring electric motors, such as streetcars systems or factories, would tend to favor DC electricity. These factors, plus the patent protection in DC systems that Edison already held, certainly supported his selection of DC-based technology for the first central station grid at Pearl Street.

Although it surely was not initially apparent in the early 1880s, one of the great limitations and weaknesses of Edison's central station grid was that the DC technology would limit the transmission of low-voltage current to about a mile and a half (about 2.5 km) from the generating station (Stana & Apse-Apsitis, 2015). (As discussed in the next chapter, workable electrical transformers, required for the transmission of electricity in either AC or DC over greater distances, had not yet been invented.) An electrical distribution system generating at a relatively low voltage of 110 volts would encounter considerable line losses[5] over any distance greater than a mile. Distribution of electricity at 110 volts required a high current, with the result that the DC distribution conduits had to be very thick to support these currents, and copper was a very expensive component in these early distribution systems (Stana

---

3 Today, North America and most of the countries in South America operates at 60 Hz, while most of the rest of the world runs at 50 Hz. Interestingly, Japan operates 60 Hz systems in the eastern part of the country, and 50 Hz in the west.

4 In fact, Edison would later dismiss the research of one his early employees into AC systems as impractical, preferring instead the simpler DC technology (Jonnes, 2004). As discussed in the next chapter, that employee, Nikola Tesla, would later turn out to be a key architect of the AC-based systems that would render obsolete so many of Edison's DC-based systems.

5 See the Glossary for a definition of "line loss."

& Apse-Apsitis, 2015). Even with these thick distribution cables there were significant line losses, and beyond about one and a half miles (about two and a half kilometers), the voltage drop along the DC feeders emanating from Pearl Street lines was so large that it was no longer possible to turn on a light bulb (Stana & Apse-Apsitis, 2015). The reach of the DC system was later increased with the development of an innovative three-wire distribution system, but this still could not extend its reach beyond about 6 miles (Millard, 1992).

Edison's vision of his future business model was consistent with the constraints imposed by this DC technology. Because of the distance limitation in distributing DC electricity, Edison's initial vision foresaw cities with power stations located at 1-mile intervals with each station serving its local grid, somewhat like the microgrids of today might appear if deployed on a large scale. For example, in early plans to service the area of Manhattan south of Central Park, one scenario saw the construction of thirty-six Edison power stations, each serving no more than roughly 3 square miles (Cunningham, 2015), and each fed with piped fresh water for the steam engines and with continuous horse-drawn shipments along city streets delivering coal and removing ash. Fortunately, for traffic congestion and air and water quality in our cities, this was not to be (Bakke, 2016).

A further factor constraining the scalability of the business model was that a diversity of customers could not easily be accommodated on the same electrical system, since the voltage of the DC current could not easily be altered after it had been generated at the dynamo. While incandescent lights operated at 110 volts, streetcars might require 500 volts, and industrial factory motors might require 1,200 volts (Bakke, 2016). This makes sense. One would not expect to run an industrial motor at the same voltage and on the same circuit as a light bulb. However, the result was that competing electrical grids were often built in the same geographic area, each serving different customers with different voltage needs. Or customers would choose to keep their private systems to meet their specific requirements (Bakke, 2016). Under either circumstance, the opportunity to share distribution systems between different customer types was lost, and the potential efficiencies of shared infrastructure would remain unmet until future iterations of electrical distribution technology, and the advent of new business models.

A New Dominant Model – Central Station Grid

The dominant business model for the "central station grid" that emerged from the Pearl Street Station, reflected the following characteristics, as represented conceptually in Figure 4.3.

Value Offering

Instead of a value offering defined by the sale of electrical equipment, Edison offered *electricity* to firms and residences. It represented the first business model based on delivery of a standardized electrical service to a group of customers from a single plant. Although Edison initially argued that the electricity in each localized central grid should be customized for specific customers (e.g., electricity for factories) or group of customers (e.g., lighting in New York's financial district), he pulled back from this distinction at the urging of investors. Instead, the grids would try to serve many different but compatible applications with the same product (Granovetter & McGuire, 1998).

Customer Identification

The previous business model defined customers as factories large enough, or individuals wealthy enough, to be able to afford a dedicated power plant. With the new business model, even a relatively small customer might be able to afford the installation of one or two light bulbs if they had proximity to a generating station. However, because of the limited range of DC distribution due to the high rate of line losses, Edison's low-voltage DC systems were located in densely populated cities such as Manhattan, with many customers close to the generation plant (Skrabec & Skrabec, 2007).

First Dominant Business Model
Localized Production

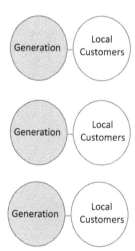

Figure 4.3: First Dominant Business Model.
Source: Author

Value Chain

Instead of being defined primarily by the manufacture, installation, and support of equipment at the customers' premises, the value chain was altered and extended to include centralized generation of electricity, distribution of that same electricity, and the provision of light to a customer through an Edison-designed light bulb. The early technologies supporting this value chain were fragmented and immature, and system efficiency was often very low by the standards of later electrical systems because the plant was tied to the consumption pattern of just a few customers.

Value Capture

Rather than follow established revenue models built around the sale and installation of electrical equipment and replacement parts, Edison would instead charge for the delivery of electricity. However, although Edison believed that electricity being sold should be measured like gas or water, the technology to measure the consumption of electricity accurately and efficiently had not yet been developed (Dyer & Martin, 1910). In fact, customers were not actually charged anything for the first several months of operation of Pearl Street (Hargadon & Douglas, 2001). However, once customer billing did start, charges in these early days were calculated "by the light bulb," while later charging schemes were based on the duration of a connection, or a flat fee based on building size regardless of actual consumption (Wasik, 2006). To measure electricity consumption more accurately, Edison developed the "chemical meter," which required, at the start and end of a billing period, weighing the amount of zinc plate deposited on a metal electrode set in a chemical solution. The customer's bill was then calculated based on the change in weight of the metal plate which was, in theory, related to the consumption of electricity in the period (Skrabec & Skrabec, 2007). By 1888, roughly 75% of the production at twenty-three of Edison's central stations was measured using this method. However, these chemical meters were found to be cumbersome and inaccurate, and prone to freezing in cold weather. Furthermore, since consumption could not be monitored or verified by the customer, they failed to garner customer confidence (Dyer & Martin, 1910). Nevertheless, chemical meters would stay in widespread use in North America and Europe until the development of mechanical meters by Elihu Thomson (Dyer & Martin, 1910) and Oliver Shallenberger (Coltman, 2002) later in the decade.

Although there were many new technological elements required for this new distribution system, the most innovative part of the business model (in this author's opinion) was the value offering. The concept of selling electricity, rather than electrical equipment, was innovative and drove major changes in all other elements of the business model. The resultant business model was totally new, and from the perspective of a customer or employee on the ground, must have been incredibly exciting.

Expansion of the Business Model

Edison's concept of a business model based around a DC-based central station grid did have a period of substantial success. Edison would utilize a licensing system to install his low-voltage DC-based systems in cities and municipalities throughout the United States, Canada, Japan, Europe, and South America over the next decade. As we discuss in the following chapters, Edison's system would eventually be supplanted by business models that we recognize today, structured on AC-based central station grid technology and a robust regulatory structure. However, these current business models certainly draw their lineage back to Pearl Street.

Nevertheless, this business model and its descendants did not emerge into dominance overnight. One of the factors that may have initially caused some early potential customers and investors to hesitate from committing to the central station grid, was an abundance of scepticism from the scientific and business community. In 1878, only months after he had laid out his vision for the delivery of incandescent lighting from a central station, a British Parliamentary Committee of Inquiry, after consultations with that country's most eminent physicists and scientists, concluded that Mr. Edison displayed "the most airy ignorance of the fundamental principles both of electricity and dynamics" (Conot, 1979, pp. 129–133, as cited in Hargadon & Douglas, 2001), and that production of incandescent lighting was not commercially feasible. Many American experts were similarly doubtful, with one proclaiming that Edison's plans for a central station were "so manifestly absurd as to indicate a positive want of knowledge of the electrical circuit and the principle governing the construction of electric machines" (Conot, 1979, p. 162, as cited in Hargadon & Douglas, 2001).

Another factor impacting the expansion of the new business model was the strong competitive response from gas companies. In 1884, two years after the opening of Pearl Street, six New York gas companies merged to form Consolidated Gas of New York. The following year, the consolidated entity dropped gas prices to $1.05 per thousand cubic feet, down from as much as $2.50 in 1860 (Hargadon & Douglas, 2001). Although lighting from Edison's electricity grid may have been cheaper than gas when it was first launched, at those new lower prices, by 1885 lighting by gas was once again cheaper than grid-supplied electricity (Hargadon & Douglas, 2001).

After opening the Pearl Street station, Edison continued to service businesses and individuals needing a supply of electricity with two parallel business models: the legacy "private plant" model (discussed in the prior chapter) and the new "central station grid" model (see Table 4.1). Of the two business models, it was not predestined that the latter should become the dominant business model. In fact, Edison's investors encouraged him not to put all his eggs in the "central station" basket and to continue the development and marketing of private plants (Bakke, 2016). Private plants provided Edison with his largest source of revenue and profitability for many years after opening Pearl Street (Jenkins, 1982). In the five-year period following the opening of Pearl Street in 1882, Edison licensed 121 central station grids, but sold over 1,200 private plants, a ratio of ten-to-one private plants to central station grids (Bakke, 2016). In many communities, private plants enjoyed an incumbent's advantage, and as late as 1915, thirty-three years after Pearl Street started operations, most electricity consumers in the United States were still served by private plants (Granovetter & McGuire, 1998). However, when change did come, it came rapidly. By 1925, over three quarters of the electricity consumed in the United States finally would come from a central station grid, built upon the business model framework that Edison had established forty years before (Bakke, 2016).

In retrospect, there is no doubt that the launch of the Pearl Street grid on 4 September 1882 represented a landmark business model innovation with far reaching implications. However, it is also true that momentous events may not always be recognized as such at the time. Interestingly, although *The New York Times* was one of Pearl Street's inaugural customers, the newspaper's first-hand account of the plant's opening was not given status as front page news but was instead filed the following day under "Miscellaneous City News" (Sulzberger, 2010). However, the *New York Herald* gave the opening of Pearl Street station greater prominence and approval, with an anonymous reporter recording the event on 5 September 1882, the day after the opening:

> In stores and business places throughout the lower quarter of the city there was a strange glow last night. The dim flicker of gas, often subdued and debilitated by grim and uncleanly globes, was supplanted by a steady glare, bright and mellow, which illuminated interiors and shone through windows fixed and unwavering. (Rutgers, 2020d)

Pearl Street operated until partially destroyed by fire in January 1890. The station was rebuilt and operated until 1894, when it was decommissioned and dismantled. Today, this historic site is occupied by an office tower, with a retail bank at street level.

Table 4.1: Summary of First Dominant Utility Business Model – 1882 to 1890s.

Business Model Element	Description
Key Characteristics	– The first electric grid. – Localized production primarily focused on lighting. – Competition against incumbent private plants and gas lighting.
Customer Identification	– Customers could now be small enough to afford the installation one or two light bulbs rather than a dedicated power plant. – Constraints of DC technology required customers to be limited by proximity to generating station.
Value Offering	– The value offering represented the sale of electricity and lighting rather than the sale of equipment, a first in North America. – Driven by Edison's vision of cities electrified by a network of DC-based power-plants.
Value Chain	– Edison's system saw all substantial elements of the generation and distribution system, down to the very light bulb itself, designed and manufactured by Edison. However, the key deliverable of the value chain was electricity and light.
Value Capture	– Metering technology to accurately measure consumption had not been developed. – Early attempt at volumetric pricing, including charges "by the light bulb" or by duration. Also, early attempts at flat rate pricing. – Later attempts to use chemical meters to estimate consumption proved to be problematic.

5 The Second Utility Model – Centralized with Competition

> My mind was upon supplying all the energy in centres of population. I realized that this could only be obtained from highly economic power stations resulting in a very low cost of energy, competing against privately owned uneconomical steam plants. The opportunity to get this large power business was right at my threshold, and I knew that unless I built the most economic power station possible, that opportunity would be lost.
>
> Samuel Insull, CEO of Chicago Edison, reporting to his board of Directors, 1902 (Wasik, 2006, p. 84)

The Emergence of AC Technology

When Thomas Edison met with his investors in 1885, the business model that he had launched three years earlier, based on the central station grid, was surely measured as a success.

From a customer's perspective, access to the new electrical grid delivered significant value, both from the benefits of electricity and incandescent light, and from cost savings. For example, when the *New York Herald* connected to the Pearl Street grid in early 1883, the newspaper eliminated gas lighting that was both unsuitably dim and a considerable fire risk in an environment stacked with newsprint (Wasik, 2006). In addition, Edison's managers estimated that they would cut the Herald's lighting costs in its first year from $20,000 per year spent on gas, to $8,000 for incandescent light supplied off the Pearl Street grid (Wasik, 2006). Cost savings would have accrued to many customers in the first few years they connected to the Pearl Street station, but this initial cost advantage over gas lighting was mitigated, and eventually eliminated in New York, by two factors. First, in response to the challenge from grid-supplied electricity, New York's six gas companies consolidated in 1884 to a single utility, Consolidated Gas, and then proceeded to slash prices (Hargadon & Douglas, 2001). Second, gas companies introduced an innovation in 1885 called the "Welsbach mantle," essentially an asbestos pouch that fit over the gas flame, providing a six-fold increase in candlepower (Hargadon & Douglas, 2001). By 1885, electricity no longer provided lower cost lighting in New York than gas (Hargadon & Douglas, 2001). Nevertheless, Edison's customer base continued to grow throughout this period, attracted to the functionality of electricity and incandescent lighting. Pearl Street's original 85 customers and 400 light bulbs (Noberni, 1982) expanded within a year to 513 customers with 10,000 light bulbs, which was effectively the plant's capacity when built (Sulzberger, 2010), and Edison would go on to build similar low-voltage central station grids in other parts of New York City (Sulzberger, 2010). Pearl Street delivered electricity to its customers continuously from its September 1882 launch until badly damaged by a fire

https://doi.org/10.1515/9783110714036-007

in January 1890, with only a single three-hour interruption of service, a remarkable record of reliability for the new enterprise (Sulzberger, 2010).

A successful business model must not only deliver value to customers, but also requires financial success for its investors. This may have been a concern to Edison and his investors in the first years of operations. With customers largely using only morning and evening lighting, the Pearl Street plant had a poor associated capacity factor,[1] and would fail to recover its operating or capital costs in the first two years of operations (Cunningham, 2015). However, in 1884, Frank Sprague, an American naval officer and inventor, developed a practical DC motor that was approved for the use on the Edison DC system (Martin, 1911). The adoption of this motor by factories significantly increased daytime customer load on Edison's New York electric system. Following losses in the first two quarters of 1884, Edison Illuminating Company declared a 6% profit for the third quarter of 1884 (Hargadon & Douglas, 2001) and turned profitable for the year in 1885 (Cunningham, 2015). Edison would go on to install 121 central station plants across North America, Japan, South America, and Europe in the five-year period following the opening of Pearl Street (Bakke, 2016). The remnants of those original franchises are found today in the names of many utilities still operating in the United States, including Ohio Edison, Consolidated Edison, Southern California Edison, ComEd, and many more.

With these measures of customer value and financial performance, and with an absence of superior competing technologies in its early years of operation, Thomas Edison's business model for a DC-based electrical grid had proven to be successful. However, advances in technology would soon expose an inherent weakness in Pearl Street's business model: a limited ability to expand the footprint of the plant's service area.

The primary use of the electricity distributed from Pearl Street was Edison's incandescent lighting, and the voltage most suited for this usage was relatively low, at about 110 volts. This low-voltage DC generation and distribution system was inherently limited in the geographic footprint it could service, since line losses in the transmission of the electricity technically limited distribution to a range of about a mile and a half (about 2.5 km) from the point of generation (later expanded through technical innovation to about 6 miles from the point of generation). Consequently, Edison's DC central stations were only economic in dense urban areas with an abundance of customer load near the station. In addition, practical electrical transformers[2] had not yet been developed that would allow voltage to be easily altered once the electricity had been generated and fed into the grid. Accordingly, customers with different voltage requirements,

---

1 Capacity factor is a measure of efficiency in an electrical system, and by most definitions is the same as "load factor." See the Glossary for further definition of "capacity factor."
2 An electrical transformer is device that raises or lowers the voltage of an electrical current. See the Glossary for a further description of "electrical transformer."

such as offices requiring lighting at 110 volts, and companies running large industrial motors at much higher voltage, could not be serviced on the same DC circuit.

Researchers and engineers did understand that that higher voltage AC and DC systems could transmit electricity a much greater distance than the 1.5 mile limitation of the Pearl Street station. As the German physicist, Georg Ohm, discovered in the early 1830s, electrical power is proportional to the product of current and voltage.[3] By increasing the voltage in a circuit, the same amount of electricity can be transmitted over that circuit, but with less current. Reduction of current in a circuit reduces the heat created in that circuit and reduces the resultant line-losses that occur with the transmission of electricity. In addition to the benefit of reduced line-loss, a reduction in current also allows smaller diameter cables to be used for distribution of electricity, and with that, a reduction in the size of structures that carry the cables. The impact of a cost reduction in the cost of cables could be significant. When the Pearl Street grid was constructed in 1882, the distribution system was the most expensive component of the entire project, in large part due to the heavy copper feeder cables required to carry the low-voltage DC current (Sulzberger, 2010).

Although practical methods for transforming the voltage of a DC current would not be developed for several decades after the opening of Pearl Street, early devices to covert AC electricity from one voltage to another had existed in the workshops of inventors and researchers for decades prior to the 1880s. However, the first truly practical AC transformer would be developed between 1883 and 1885 at the Hungarian manufacturing firm Ganz Works by a team of three engineers, Otto Blathy, Miksa Deri, and Karoly Zipernowsky. Their invention became known as the "ZBD transformer" (named after the inventors), and by the fall of 1886 there were early electrical systems using the ZBD transformer in several European cities (Allerhand, 2019). However, any attempts to implement the ZBD transformer technology in the United States would require the cooperation of Edison Electric, which had purchased an option on the rights to this system for $5,000 (Skrabec & Skrabec, 2007). Edison, of course, was intent on protecting his own DC patents, frustrating entrepreneurs like George Westinghouse (see Figure 5.1) who sought to utilize the AC technology.

Westinghouse responded by commissioning the American engineer and inventor, William Stanley, to develop his own home-grown, improved, electrical transformer technology. In 1886 Westinghouse and Stanley did produce an AC transformer based on technology that was differentiated from the ZBD transformers, allowing them to secure their own patents. This was followed in 1887 with the invention of a workable AC motor by Nikola Tesla (see Figure 5.2) and other technologies that would support the implementation of an AC electrical system (Roguin, 2004). Westinghouse travelled to

---

3 In mathematical terms, this is expressed as ($P = I \times V$) where P is power (measured in watts), I is current (measured in amperes), and V is electrical potential or voltage (measured in volts).

Figure 5.1: George Westinghouse, circa 1884, Age 38.
Source: Photograph by unknown author, distributed on Wikimedia Commons under a CC-BY 4.0 license.
https://commons.wikimedia.org/wiki/File:George_Westinghouse_1884.png

Tesla's laboratory, not far from Edison's own Menlo Park laboratories, and struck a deal to purchase a portfolio of forty AC patents for cash and stock worth $60,000, plus royalties on future electrical capacity sold (Roguin, 2004). With the addition of Tesla's patents, the technological pieces that Westinghouse needed to compete with Edison's DC system were taking shape. Westinghouse' newly designed AC system would generate electricity at 500 volts, and then using William Stanley's transformers it would step up the current for transmission for some distance at 3,000 volts, before stepping back down to lower voltage for distribution to customers (Bakke, 2016). Edison's Pearl Street station, by comparison, originally generated and distributed electricity at only 110 volts, a voltage most suitable for incandescent lighting (Sulzberger, 2010), but ill suited for transportation over longer distances. Although the DC voltage at Pearl Street would later be doubled to 220 volts to reduce both line losses and the amount of copper needed for conductors (Sulzberger, 2010), without the availability of practical DC transformers to increase or decrease voltage after generation, low-voltage DC current would continue to be limited in the distance it could be distributed.

Figure 5.2: Nikola Tesla, 1890, Age 34.
Source: Photograph by unknown author, distributed on Wikimedia Commons under a CC-BY 4.0 license.
https://commons.wikimedia.org/wiki/File:Tesla_circa_1890.jpeg

Although the low-voltage output of Edison's Pearl Street Station in 1882 only had a range of about 1.5 miles (about 2.5 km), DC technology did evolve such that it was used in 1889 to transmit high-voltage DC electricity for 13 miles (21 km) from the Willamette Falls hydroelectric plant to Portland, Oregon, the first long distance transmission of electricity in the United States (Nichols, 2003). The technology to transmit either AC or DC electricity over distances of more than a few miles was unproven at that time, and its immaturity was demonstrated in that the use of electricity was only one of four alternatives described by a Swiss engineer who had been commissioned in 1884 to examine the transport of energy from the Willamette Falls to Portland. The engineer described four options: the use of wire rope on pulleys, compressed air, water at high pressure, or electricity (Nichols, 2003). The relative merits of the competing technologies at that time, and the unproven nature of electricity as a means of transporting energy, are indicated in the consulting engineer's comments: "The cheapest and most effective way to transmit hundreds of horsepower from one place to another is by means of endless wire ropes running over large pulleys" (Nichols, 2003, p. 8).

Fortunately for the city of Portland, the investors in the scheme a few years later chose electricity as the means to transmit the energy of the falls. Belt-driven dynamos were repurposed from a local sawmill to generate DC electricity, and by June of 1889 the streets of Portland were brightened with electric lighting energized from the falls (Nichols, 2003). However, by the following year, 1890, this newly constructed DC transmission system to Portland had been deemed inadequate, and the DC dynamos were replaced with six Westinghouse AC generators, feeding what was then the first commercial transmission of AC power in the United States. The immaturity and newness of using this technology to transmit electricity over long distances was demonstrated in an initial hesitation of Westinghouse to supply the AC generators for such a task, but eventually agreeing to do so only with the "express condition that the company was not to be held responsible for and would not undertake to guarantee satisfactory operation of such generators" (Nichols, 2003, p. 9).

Westinghouse continued to increase the reach of its AC transmission capability with an 1892 installation in California of a 28 mile (45 km) AC transmission line from the San Antonio Canyon hydroelectric plant to San Bernardino, and a second 14 mile AC line to Ponoma (Allerhand, 2017). Meanwhile, in Europe, AC technology was also being used to transport electricity over longer distances, with the installation in 1889 of 28 miles of transmission line from a new coal-powered station in Deptford to Greater London, using designs by Sebastian de Ferranti, a British engineer (Guarnieri, 2013). In 1891, the Electro-Technical Exhibition in Frankfurt demonstrated the first experimental three-phase AC transmission line, carrying electricity 109 miles from the Lauffen waterfalls on the Neckar River (Guarnieri, 2013). This line was not put into commercial use, but it did lead to the world's first commercial three-phase AC transmission line from the Lauffen Falls to Heilbronn, Germany in 1891, although with a distance of only 6 miles (Allerhand, 2017). This was followed in 1892 with a 17-mile AC line utilizing the ZBD transformers developed by Ganz Works to connect Rome with a hydroelectric plant at the Aniene Falls in Tivoli (Guarnieri, 2013). These projects (see Table 5.1) were demonstrating the advantages offered by the ability to transform AC current to high-voltage for transportation, and stepping voltage back down for distribution, a capability that DC-based technology could not yet match. This capability opened the way for a new business model, based on locating power stations close to their primary energy source (e.g., close to a hydroelectric source, or a transportation hub for coal supply) and using transmission over longer distances to deliver electricity to centers of demand (Guarnieri, 2013).

Table 5.1: Notable Transmission Installations.

Period	Description
1889	Willamette Falls to Portland––13 miles (DC)
1889	Deptford Power Station to Greater London––28 miles (AC)
1890	Willamette Falls to Portland––13 miles (AC)
1891	Lauffen Falls to Frankfurt––109 miles (experimental three-phase AC)
1891	Lauffen Falls to Heilbronn––6 miles (three-phase AC)
1892	Aniene Falls, Tivoli to Rome––17 miles (AC)
1892	San Antonio Canyon hydroelectric plant to San Bernardino (28 miles), and a second line to Ponoma––14 miles (AC)
1893	Selection of AC for Niagara Falls to Buffalo
1896	Niagara Falls to Buffalo in service––22 miles (AC)

Source: Nichols, 2003; Guarnieri, 2013; Bakke, 2016; Allerhand, 2017.

This period in the late 1880s and early 1890s would see an epic and very public competition between Edison and Westinghouse for dominance of AC or DC technologies, a battle that would come to be known as "The War of the Currents," and still the subject today of popular books and Hollywood movies. A key milestone in this intense rivalry took place in 1893 with the selection of Westinghouse's AC technology for North America's first large scale power plant at Niagara Falls, New York, a selection that had been preceded by years of debate in the scientific and business communities. The Niagara Falls project, which included the completion of the power plant in 1895, and a 22 mile transmission line to Buffalo, New York, in 1896 (Bakke, 2016), wasn't the first demonstration of the superior geographic reach of AC technology, but it certainly was the largest and most publicity-laden to that time. The project was called "the unrivalled engineering triumph of the nineteenth century" by *The New York Times* (Munson, 2005, p. 41), and the abundant electricity distributed through that region of the state of New York would attract the world's largest group of electrochemical companies, using the secure source of electricity to develop industrial scale manufacturing of products such as graphite and abrasives. It also attracted the firm that would eventually become Alcoa (Munson, 2005), using the secure supply of electricity to develop the first commercial, large-scale electro-processing of aluminum (Bakke, 2016). Within years, Niagara's electricity was transported all the way to New York to run the city's new subways and to light Broadway (Munson, 2005).

The "War of the Currents," had largely been won by Westinghouse's AC technology by 1893. It would no longer be necessary to situate generation in close proximity to the customer, and by 1894, the year after the technology for Niagara Falls had been selected, 80% of new electricity grids in the United States were being designed with AC technology (Bakke, 2016). Electrified North America migrated from Edison's DC electrical systems and toward a value chain that delivered AC electricity using the standards first established at Niagara Falls: alternating current, oscillating at 60 Hz, transmitted at high voltage, stepped down to lower voltage and then distributed to homes and businesses at 110 or 220 volts (Bakke, 2016). In terms of technical standards, this established the grid that North America still has today.

A little more than a decade after its introduction, Edison's business model vision of "central station grids" dotting cities at square-mile intervals was largely rendered obsolete by the emergence of this new AC technology. The AC technology enabled an entirely new business model with a more diverse customer base, much larger scale economies, and later, the ability to use new rate structures to alter consumption patterns that would further lower costs. If the DC-based business model had survived, perhaps the grids would have resembled distant cousins of today's microgrids: a relatively small distribution network with local production, perhaps attached to a larger grid. But relative to the widely adopted AC systems, Edison's DC-based grids could not have achieved the scale economies and massive cost reductions achieved by AC-based systems.

It is notable that technological advances in semiconductor devices now allow DC power to be transformed to high voltage, decades too late to save Edison's DC-based business model. In fact, as discussed later in this chapter, high-voltage DC (HVDC) is now often the most cost-effective mode of electricity transmission for distances exceeding 500 miles (about 800 km) (Thomas, Azevedo, & Morgan, 2012). With today's technologies, AC remains the norm for local transmission and distribution, although the considerable research being undertaken into HVDC may also change those economics in the future.

A Challenge to the DC-Based Business Model

With the advent of the Westinghouse AC grid, a new dominant business model emerged. With the ability to transmit electricity over greater distances using alternating current, firms were able to develop larger service territories, thereby accessing a wider range of customers. With this larger and more diverse customer base came the opportunity to pursue scale economies and reduced costs. One of the first to recognize this potential was Sam Insull, a former right-hand man of

Thomas Edison who, ten years after helping to open Edison's Pearl Street station, became president of Chicago Edison[4] (Wasik, 2006). The expansion of this AC-based central grid had started throughout the industrialized world at that time, and Insull's work in growing Chicago Edison is not the only example of this growth. However, Chicago Edison is one of the best-known and successful examples of this transition to a new business model and formed a successful template for others to follow.

When Insull arrived at Chicago Edison in 1892, the company was only one of eighteen central station electricity providers within Chicago's downtown loop, which was also home to an additional 500 private power plants (Bakke, 2016). Chicago Edison in 1892 operated two parallel business models. First, the company had a successful business of installing private stations, largely to industrial customers. Second, Chicago Edison had installed the city's first central station grid in 1888, supplying electricity to approximately five thousand customers (Bakke, 2016). However, since this system was based on Edison's DC technology, it was inherently limited in the geographic area that the plant could serve. In the decade following his arrival at Chicago Edison, Insull recognized the opportunity to massively grow the company's central station business, while effectively relegating the company's private station business to the sidelines. The next few pages will discuss some of those actions that transformed the business model Insull had inherited.

Consolidation and Investment in New Technology

In this transition to a new dominant business model, Insull would totally restructure the value chain for the Chicago Edison central station business. By 1897, five years after Insull's arrival as CEO, Chicago Edison had taken over or put out of business every other central station utility in central Chicago, and in doing so had also acquired a portfolio of small, inefficient, generating stations (Wasik, 2006). Insull started a program of investment in replacing these small stations with ever larger plants. By 1894, Chicago Edison had closed its first 3.2 MW Edison generating station, built only six years before, and opened a new 6.4 MW station at Harrison Street, eventually expanding its capacity to 16.4 MW (Wasik, 2006). The new station was the largest in the world when constructed and produced electricity at half the unit cost of the station it replaced (Lambert, 2015).

The construction of larger generation plants, such as the Harrison Street plant, were initially constrained by the requirement to service existing local distribution systems that had been built to operate with DC electricity (Bradley, 2011). This was a problem, since AC power was needed to transmit the output of these large plant

---

4 Chicago Edison merged with Commonwealth Electric in 1907 to become Commonwealth Edison, a predecessor company to today's ComEd that still services the Chicago area.

over greater distances to a larger customer base, but many of these legacy local networks needed to be served with DC. Fortunately, technology presented a solution to this constraint, as a newly developed rotary voltage converter could be installed at a substation some distance from the station but close to the consumer, where it could step down high-voltage AC to low-voltage DC for distribution to homes (Bradley, 2011). With this new conversion technology, the Harrison Street station's reach was greatly extended. The Harrison Street plant was followed only nine years later, in 1903, by a four-times larger AC station at Fisk Street (see Figure 5.3), with 68 MW plant capacity, equipped with state-of-the-art 5 MW General Electric steam turbines that each produced twice as much power as any previous steam engine ever built (Wasik, 2006). Enabled by fierce completion between GE, Westinghouse, and Siemens, Consolidated Edison was installing 24 MW steam turbines by 1910 and 120 MW turbines by 1920 (Wasik, 2006). The growth in capacity was remarkable.

Figure 5.3: Fisk Street Station, Generator Unit, circa 1903.
Source: Library of Congress, Prints & Photographs Division, Call number: HAER ILL, 16-CHIG, 140–16 HABS
Library of Congress. http://www.loc.gov/pictures/item/il0671.photos.034796p/

As generation plants grew larger and benefited from technological innovation, such as the shift from reciprocating steam engines to steam turbines, so too did their thermal efficiency (i.e., the measure of the rate at which the electric plant converts the energy potential of the fuel, such as coal, into electric energy). Pearl Street's thermal efficiency in 1882 ran at about 2%. By the time Insull opened Fisk Street in 1903, the plant efficiency was well over 12% (Bakke, 2016). (By comparison coal-fired generation plants operated at about 20% in the 1940s [Bakke, 2016] and today typically run at about 38% efficiency [Smil, 2010].) These improvements in plant efficiency would be the first building block to an incredible improvement in cost reduction and profitability. The next would be a much different identification of the customer.

Diversification and Growth of the Customer Base

Imagine the load profile of a typical Edison DC utility in the early 1880s, that primarily delivers lighting to businesses in a dense, city center. During the day, those businesses that could operate on natural light, would do so. (Electric lighting during the day would still be expensive for most businesses.) As dusk fell, they would turn on their electric lighting to allow work to continue until closing time at 5 or 6 p.m. In the evening, as employees went home, electricity consumption in those businesses would drop to very low levels. Workers went to their homes in the suburbs, but because electric services had not yet extended to less densely populated areas, they would probably light their homes with kerosene or gas, and not electricity (Bakke, 2016). Although the grid's mid-day load was improved by businesses installing Sprague's DC motor after 1884, the typical central station power plant would still have a very low utilization for much of the day. When Sam Insull arrived in 1892, this low utilization, or "capacity factor," is exactly the situation he found facing Chicago Edison, a not unusual situation for Edison central station franchises. With its DC central station plant aimed at servicing approximately 5,000 lighting customers in fifty-six city blocks in the city's business core (Wasik, 2006), Chicago Edison's 3,200 kw generating plant would sit underutilized for much of the day, after customer's offices closed and demand for lighting ceased. As Insull, noted, "If your plant is only in use 5.5% of the time, it is only a question of when you will be in the hands of a receiver" (Bakke, 2016, p. 65).

Matching plant capacity to customer load is a problem that challenges electrical utilities to this day, and is common across many industries with high fixed costs, like pipelines or airlines. To address this problem, Insull needed to develop a more balanced customer base; he needed to find customer load outside of the hours in which businesses needed late afternoon lighting. Over the next few years, Chicago Edison expanded into servicing not only local lighting customers, but also factories requiring electric motors running during the day, and streetcar companies running from morning to evening. Over the next few years, as power systems switched to AC and the geographic footprint available to the power plant extended into the suburbs, the customer load expanded to include residential customers needing morning and evening lighting and, later, additional electricity for newly acquired appliances (Bakke 2016). With these new and diverse sources of customer load, utilization of the utility's electrical system would be markedly higher, increasing the system capacity factor and driving down unit costs. In addition, in 1893, Chicago Edison started to purchase other central stations companies, integrating their service territories. By 1906, Insull had grown Chicago Edison's customer base tenfold, to 50,000 customers, and by 1913 the company reached 200,000 customers (Bakke, 2016).

Reduction in Costs and Prices

It is a bold act for an enterprise to expand a new, unprofitable business model to cannibalize an existing, profitable one. Modern day examples might include Netflix's use of electronic media delivery to successfully drive its postal-delivery, DVD rental business to obsolescence. In Chicago Edison's case, as typical for many Edison franchises in 1892, the sale and maintenance of private plants formed the most profitable part of its business, while the operations of its central station grids were much less profitable (Bakke, 2016). Over the next few years, Insull would achieve massive cost reductions and growth in the central station grid business, resulting in the business model focused on private plants occupying a small and diminished share of Chicago Edison's total business (Wasik, 2006).

Two key technological breakthroughs enabled Insull to achieve remarkable unit cost reductions on Chicago Edison's central station grid. First, introduction of AC technology greatly expanded the geographic footprint available to be serviced by each electrical plant, increasing scale and customer diversity, thereby increasing system efficiencies, and driving down unit costs. Second, the shift from steam reciprocating generation to steam turbine technology greatly increased generating plant capacity while also reducing unit costs. Driven by a dramatic increase in plant efficiencies, and a remarkable reduction in unit costs, the company slashed electricity prices to the point that drove industrial customers away from the use of private stations, and onto Chicago Edison's central station grid. From a starting point of 20 cents per kWh when he arrived in 1892, Insull cut the price of electricity by half, to 10 cents per kWh by 1897, and in annual increments to 2.5 cents per kWh by 1909, a reduction of 88% in 17 years. To attract night-time load when his system was otherwise largely idle, Insull actually cut prices in 1911 to only 0.5 cents per kWh for off-peak industrial customers (Bakke, 2016). Growth of customers on its central station grids took off, and with addition of generation capacity at its Fisk Street station in 1903, Chicago Edison could use AC transmission systems to extend its reach outside the city, allowing further consolidation of small utilities and displacing private systems. In 1910, Insull used Lakeland County, Illinois as a test bed to show the benefits of consolidation of several small utilities that existed in the county at that time (Kahn, 1984). In an address to the Franklin Institute in 1913, Insull demonstrated that between 1910 and 1912, by unifying Lakeland County's multiple utilities and doubling the number of customers served, he was able to cut prices in the community by 18%, from 9 cents to 7.25 cents per kWh (Kahn, 1984). Insull had demonstrated it: consolidation worked.

Improved Volumetric Pricing

Revenue capture by utilities had been a difficult problem for the first generation of central station grids. While a utility could closely manage its costs, it had limited means of measuring how much product had been consumed by each customer. Some of Edison's early Pearl Street customers were charged on a per-light-bulb basis, others based on building size, or charged flat fees (Wasik, 2006). Later attempts to measure consumption with chemical meters were regarded as imprecise and not widely accepted by customers. Eventually Edison did develop a motor-based mechanical meter that worked for DC current, but would not operate on AC current (Ruch, 1984). Dozens of inventors worked to measure consumption of electricity more accurately, by measuring mercury in a glass tube, heating vials of alcohol, or recording the differential in dual mechanical clocks. Each design suffered shortcomings. Eventually, in 1887 and 1888, Oliver Shallenberger, an American naval engineer hired by George Westinghouse, developed a prototype of an electromechanical meter that included the technology that would define meters for more than the next century (see Figure 5.4).

Figure 5.4: Shallenberger's Meter, 1887.
Source: This image of Oliver Shallenberger's meter is taken from the 1887 publication *The Weekly Engineer* (Pope, Phelps, Martin, & Wetzler, 1887, p. 383)

Although Shallenberger's device was subsequently improved by many others, his basic design remained, using the rotational speed of a disk in an induced magnetic field of an AC electrical current to measure watt-hours (Sowmya, Kumar, & Gangadhar, 2016). Since the rotating disk in Shallenberger's design operated on the induced magnetic field, this allowed the meter to operate while consuming almost no power itself. As those of us who have watched the rotation of the disk in the meter on the side of our own homes will recognize (while wondering how the teenagers in the house can possibly cause it to spin so quickly), this formed the basic technology used in mechanical meters for over a century, until replaced by digital smart meters in the twenty first century. For the first time in North America, utilities widely deployed metering technology that could accurately charge customers based on the amount of their actual kilowatt-hour consumption of electricity (Wasik, 2006).

While the per-kilowatt-hour charge was reasonably accurate, simple, and easily understood by customers, it was an imperfect method of matching revenues with the cost of doing business. A central station grid is a capital-intensive business, and the total grid system, from generation to customer, must be built to consider the peak demand that will be placed on the system. And since most of the cost of building and financing that peak capacity is fixed, that posed a problem for many early grids with low utilization rates, or "capacity factors." The lower the capacity factor, the higher the average unit charge had to be to cover the costs of the plant. However, with high unit charges, customers would tend to consume less, and profits would be tougher to achieve (Lambert, 2015). The standard consumption-based charge enabled by Shallenberger's meter, although accurate and easy to understand, did not incent customers to increase their consumption, nor did it incent them to shift consumption to time periods when the system was not already at peak output.

A breakthrough in measurement capabilities of North American utilities came when Sam Insull, while on holiday in Brighton, England, in 1895, discovered a local installation of an innovative electro-mechanical metering system. In discussion with the inventor of the meters, Arthur Wright, who was also the manager of the local central generating plant in Brighton, Insull discovered that the meter recorded not only how much the customer used during a billing period, but also the magnitude and duration of the peak level of consumption during that period (McDonald, 2004). Wright argued that if you had two customers who used the same amount of electricity, but one customer had a much higher peak consumption, it would make no sense to charge them the same rate since they have such differing impact on the system's fixed costs. Instead, with Wright's meter, it would be possible to measure each customer's use of two different elements: the amount the customer contributed to the system's peak capacity requirements, which contributed to the requirements for total fixed costs, and the amount of hourly consumption, which contribute primarily to variable operational costs (McDonald, 2004).

With these insights Insull returned to Chicago, but would immediately send a representative back to Brighton to license the meter technology for use at Chicago Edison (Wasik, 2006), and hired Arthur Wright to come to Chicago. Using the insights gained from Arthur Wright in Brighton, Insull set to work on a new approach to customer charges: a two-part bill. The first charge on the bill would be intended to pay for the customer's share of the utility's fixed costs of delivering the service, and would largely be carried by big users. The second charge, representing the customer's consumption, would be at a much lower rate, a rate that would further decline as consumption increased (Tobin, 2012).

Insull had determined that profitability was not related only to load (i.e., the quantity of electricity sold), but more importantly was driven by capacity factor (i.e., the amount of energy a power plant actually produces, compared to the amount that it would produce if run at full capacity) (Lambert, 2015). "The nearer that you can bring your average to the maximum load, the closer you can approximate the most economical condition of production, and the lower you can afford to sell your current" (Lambert, 2015, p. 12). Insull equipped himself with a chart, showing a daily system load profile with morning and evening peaks, as customers used lighting, streetcars and the elevated railway during those morning and evening hours. Between the peak hours were, of course, valleys of low consumption (Tobin, 2012). As the historian Thomas Hughes described, referring to the charted load profile, "Insull and his associates . . . did everything possible to fill in the valleys" (Hughes, 1993 as cited in Tobin, 2012). Using the new two-part rate structure, Insull could now offer much lower rates in off-peak periods to encourage consumption. He provided deals to factories to stretch working hours to include overnight shifts and encouraged all-night restaurants to connect to the grid (Tobin, 2012). He grew the size of his sales department and started a campaign to promote the growing use of appliances in the home. Chicago Edison trucks loaded with new electric irons went into residential areas, offering free use of an iron to anyone who signed up for service from the grid (Tobin, 2012). This changed consumption patterns markedly. In the five years following 1894, Chicago Edison went from selling almost three quarters of its electricity for lighting, to almost half for small power applications. More importantly, the company doubled its capacity factor and customer base, while tripling the electricity volumes sold (McDonald, 2004). Increased capacity factors and reduced costs were good for customers as well, with the average customer bill falling by 32% in 1898 alone (Tobin, 2012).

Early Regulatory Oversight of Prices

The revenue collected by utilities during this period started to come under the first attempts at regulatory oversight by state governments, and by 1922 almost thirty states established regulatory commissions for the oversight of public utilities and passed legislation to establish municipal regulation of utilities (Troesken, 2006). Canada followed a similar pattern, with the first regulatory authority established in Ontario in 1906 and other jurisdictions following (Rosenbloom & Meadowcroft, 2014).

In these early days of regulation, "cost of service"[5] rate-setting methodologies had not yet been widely developed. Instead, most rate regulation of private sector electric companies was based on establishment of price ceilings, with only passing reference to the public utility's cost of service. Nevertheless, the first foundations of regulation had been established, and would form a foundation for the increased regulation that would be central to Insull's part in the development of the next dominant business model, discussed in the next chapter.

A New Dominant Model – Centralized Production, Fragmented Distribution

The first dominant business model implemented by Edison at Pearl Street would be constrained by the inability to transmit DC electricity any further than about 1.5 miles (about 2.5 km). The ultimate demise of Edison's business model was initiated by series of technological breakthroughs in the utility value chain. First, new steam turbine technology enabled break-through gains in efficiency and plant size, while the development of AC-based equipment, led by George Westinghouse, enabled utilities to reach a much greater customer base from a single generating station. The breakthrough in metering technology, which enabled new customer rate structures, further increased customer growth and diversity. This growth in load enabled much larger generating stations to be built, while the growth in customer load diversity could allow these plants to operate much more efficiently, with much higher capacity factors.

Observed through the four-element business model framework, Insull's business model formed a template for other vertically integrated electrical companies to follow and came to dominate the industry. This business model, represented conceptually in Figure 5.5, carried the following characteristics:

Value Chain

With the introduction of AC technology and the ability to transport electricity over long distances, the basic utility value chain was introduced that has existed for decades, of "generation-to-transmission-to-substation-to-distribution-to-customer." However, despite the huge gains in efficiency, there were still great opportunities for further efficiency gains as competing AC systems offered multiple technical and incompatible standards.

---

5 Note that "cost of service" regulation will also be discussed in the next chapter. Also, see the glossary for a short definition of "cost of service" regulation.

Second Dominant Business Model
Centralized Production, Competition

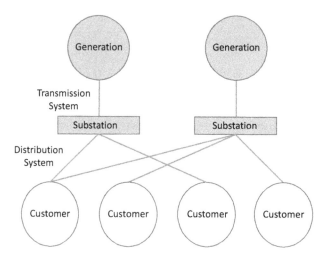

Figure 5.5: Second Dominant Business Model.
Source: Author

Customer Identification

Customers were now defined by access to transmission and distribution systems, and not by proximity to the generating station. For utilities installing AC technology, this allowed access to customers across a wider geographic area than previously available. However, utilities did not have the status of a monopoly service provider that we are familiar with today. In many markets, they were one of several competing electricity service providers, vying for customers with other utilities in the marketplace.

Value Offering

For many customers in this period of expansion, the utility's offering represented their first connection to a grid, and the opportunity to replace gas lighting and appliances with electric. For industrial and commercial customers, the value offering represented the opportunity to electrify their operations, introduce electric lighting, and replace mechanical power with electric motors. For all customers, electrification over the period was characterized by a sharp decline in the cost of electricity.

Value Capture

With the introduction of the electro-mechanical electric meter into North America, utilities were able to widely introduce volumetric pricing, based on metered, per kilowatt hour consumption of electricity. Utilities were also able to use new metering technology and new rate structures to offer volume discounts to encourage consumption, discounts for off-peak usage to increase the electric system's efficiency, and charges to larger users to reflect their share of the system's fixed cost of service delivery. Early forms of regulatory oversight were introduced but were particularly subject to political pressure and corruption.

Although advances in technology on the supply side of the business are the most visible drivers of the remarkable increases in efficiency and reductions in cost that occurred in this period, the contribution of new revenue capture tools on the demand side of the business were core to increasing load and improving load factors.

Sam Insull understood that the issue of value capture was core to the success of the new business model he had done so much to define (see Table 5.2) and had become central to the profitability of his business. As he declared, the subject of rate making had become, "the most important one that we can consider; I devote more of my time to it than to any other one subject connected with our business. The way you sell the current has more bearing on . . . cost and profit than whether you have the alternating or direct-current system, or a more economical or less economical steam plant" (Platt, 1991, para. 213).

Transition to the New Dominant Business Model

Electrification of homes and businesses took decades to roll out across much of the country. Although Edison started to provide service from Pearl Street in 1882, it was not until twenty-nine years later, in 1911, that electricity was distributed to all of Manhattan, one of the most densely populated regions in the country (Smil, 2005). In 1907, electricity still reached only 80% of American homes (Munson, 2005). Much of rural America would not start to be electrified until the Rural Electrification Act of 1936 (Bakke, 2016) and parts of Ontario and the rural Canadian prairies were still being electrified in the 1950s (Fleming, 1991). Until the grid arrived, homes and businesses relied on self-provision of electricity or energy from other sources.

If transition to electrification was neither clean nor rapid, nether was the transition to a new business model enabled by AC power. Even for those communities that had been electrified, the transition between DC-based and AC-based business models took some time to achieve. Even though AC was seen to have won "The War of the Currents" and had come to dominate the installation of new grids, the legacy DC grids continued to operate for many years, particularly in densely populated areas that that had already installed DC systems. As late as 1902, the number of DC and AC

Table 5.2: Summary of Second Dominant Utility Business Model – 1890s to 1920s.

Business Model Element	Description
Key Characteristics	Centralized production. Structured for competition.
Customer Identification	Customers were defined by access to a much larger system of AC electricity distribution, and not constrained by proximity to the generating station. Increased customer base and customer diversity allowed significant system efficiencies.
Value Offering	For many customers, the expansion of the AC network represented their first connection to a grid. Electrification accompanied by a sharp decline in unit cost. Customers with different voltage requirements could be serviced from the same grid.
Value Chain	With the introduction of AC technology, the value chain structure was introduced that still exists today, of "generation-to-transmission-to-substation-to-distribution-to-customer." Generating stations could be sited close to their primary energy source (e.g., a transportation hub for coal supply) with transmission of electricity over longer distances to larger centers of demand. Steam turbine technology enabled greatly increased plant size and efficiency, and reduced unit costs.
Value Capture	Volumetric pricing, with charges by the metered kilowatt hour. Early implementation of a form of demand charge was a step to better matching of costs to revenues. Prices established by competition, sometimes price-capped by early regulation.

central station grids in the United States were still roughly equal (see Table 5.3). Nevertheless, the tide turned over the following decade, with AC central stations growing by 40 %, while DC stations fell by a one-third (Allerhand, 2017).

The legacy DC grids often operated in a hybrid configuration with AC transmission systems. In Manhattan, for example, at the start of the twentieth century, high-voltage AC power was being supplied to sixteen "conversion substations" to be converted to low-voltage DC electricity and distributed over legacy DC distribution systems (Cunningham, 2015). At that time, 90% of the electricity delivered in Manhattan was sold this way, and even until 1915, customers on the New York Edison DC systems would continue to receive replacement light bulbs at no extra charge (Cunningham, 2015). Similar hybrid systems, with high-voltage AC generation transmitted to conversion stations for low-voltage DC distribution to customers, continued to operate in many cities that had installed Edison DC systems into the 1920s. However, building and operating the conversion substations was costly, the thick copper DC distribution cables

Table 5.3: DC and AC Central Stations in the US, 1902–1912.

Year	DC Stations	AC Stations
1902	2,607	2,634
1907	2,130	3,446
1912	1,719	3,729

Source: Allerhand, 2017, p. 776

were expensive, and by the late 1920s conversion to AC networks had commenced in many cities (Cunningham, 2015). Nevertheless, in the 1970s, DC systems were still in operations in 315 American cities (Cunningham, 2015), and some still remain. In San Francisco, for example, the electric utility still maintains a legacy DC distribution system in the downtown area to operate DC powered elevator motors in historic buildings (Fairley, 2012). In the last few years, interest has grown in the rebirth of DC-based distribution systems, driven in part by the growth in high-voltage DC (HVDC) transmission systems, and by the growth in DC-based devices attached to the grid, such as renewable energy sources and electric vehicle batteries (Wang, Goel, Liu, & Choo, 2013). The "War of the Currents" may not be over yet.

An Update on Transmission

In this chapter, the emergence of AC transmission at high voltage in the late 1880s and early 1890s is recognized as an important driver in the demise of the business model based on Edison's DC-based central station grid. However, with the development in recent years of new technologies, transformation of high-voltage DC (HVDC) has been increasingly used as the most economical and effective means of transmitting large amounts of power over a long distance, an issue of increasing importance with the growth in renewable energy generation in locations remote from load centers.

Although there are costs of converting to and from AC, HVDC transmission generally offers financial benefits from several factors at distances exceeding 500 miles (about 800 km) (Thomas, Azevedo, & Morgan, 2012). There are several factors that result in this advantage. First, HVDC tends to have lower construction costs since it typically requires infrastructure to support only one conductor, whereas an AC system would require three. Second, since there is only one conductor, cost savings can be realized from reduced right-of-way requirements. Finally, HVDC systems typically have lower line losses over long distances than AC systems (Ardelean & Minnebo, 2015).

In addition to its financial advantages in transporting large amounts of electricity over long distances, HVDC also has advantages in other situations. For example,

HVDC is used for longer distance transmission under bodies of water, since the length of today's underwater AC transmission cables can not technically exceed about 65 miles (or about 100 km) (Allerhand, 2017). In addition, HVDC is the most reliable method of connecting two AC grids operating at different frequencies. For example, HVDC connects Japan's grids, which operate at 50 Hz in the eastern part of the country, and 60 Hz in the west (Ardelean & Minnebo, 2015).

The first modern, commercial HVDC transmission line was a 60 mile (98 km), underwater 100 kV, 20 MW system, constructed off the coast of Sweden in 1954 (Ardelean & Minnebo, 2015). Today, the longest HVDC transmission lines stretch for thousands of miles, such as the 2,000 mile (3,300 km), 1100 kV Zhundong–Wannan system commissioned in China in 2019 (Saqib, 2019). The world has growing requirements to connect renewable electrical wind and solar resources in remote areas to centers of demand. Additional plans for dozens of long distance HVDC transmission lines are on drawing boards across the world to transport, for example, electricity from the wind farms of Oklahoma to cities of the eastern United States, or from the planned, massive solar farms of North Africa to the European grid (Saqib, 2019).

On Crediting Technological Innovation

The history of the development of electrical technologies is a long and winding road, with many inventors contributing to each new technology, often independent of each other. In the investigations of the 1880s and 1890s into electrical generation and distribution, researchers in Europe and America often worked on the same problems at the same time, sometimes without knowledge of one another, but sometimes exchanging results and building on each others' findings. For example, William Stanley, an American physicist, is often credited as the developer of the first workable AC transformer in the United States in 1895. However, Stanley was building on the research of a trio of Hungarian developers, Otto Blathy, Miksa Deri, and Karoly Zipernowsky who developed an earlier AC transformer in 1883 (Jeszenszky, 1996). And these three inventors leveraged the AC transformer prototypes developed by Lucien Gaulard and John Gibbs in 1882 (Jeszenszky, 1996) And these, in turn, leveraged principles of induction developed by Michael Faraday in the 1830s. And there were many others who contributed to the development of the technology. In summary, there is no true "parent" of the AC transformer, or the electrical meter, or the light bulb, or of so much of what has led to today's electrical grid. Rather, there is a community of researchers, inventors and tinkerers that stretch back decades.

This is a book that discusses the evolution of business models. As such, it uses the history of the development of electrical technologies to understand the business models that result, but it is certainly not a book written by a historian. In this book, a single person is sometimes referenced in relation to an innovation. For example, Stanley is noted as the developer of the first workable AC transformer. This makes for an easier narrative to describe the development of the business model, but it is not intended be a complete description of a complex history. If readers are interested in the history of the development of electrical technology, then they are urged to use the Reference section to discover many fantastic writings about the remarkably interesting people who occupy it.

6 The Third Utility Model – Regulated Monopoly

> Our business is a natural monopoly. It must of necessity be regulated by some form of governmental authority."
>
> Samuel Insull, CEO of Chicago Edison, 1897 speech to the National
> Electric Light Association (Wasik, 2006, p. 79)

The Mess of Competition

In the four decades following Edison's opening of his Pearl Street Station in 1882, the new electric utility industry went through an incredible transformation. Think about the changes in these few years in the four elements in the business model framework: the value offering, value capture, value chain, and customer identification. The *value offering* of grid-supplied electricity allowed customers to retire steam-based and water-based mechanical power at thousands of factories, replaced by the electric motor, while incandescent lighting replaced gas lights and kerosene lamps in millions of homes and businesses. Electrical metering enabled utilities to implement *value capture* mechanisms that tied revenues to the costs of production and enabled new rate structures that could greatly improve load factors. The *value chain* was transformed by the efficiency of steam turbine technology and the reach of AC transmission to enable distribution of electricity at much lower cost over an increasingly large area. This, in turn, reframed *customer identification* to be defined not by proximity to a generating plant, but rather by proximity to the much larger grid, and allowed utilities to target customer consumption that would greatly increase overall system efficiency.

Despite the success of the growth of electricity as a product during this period, the business model that supported this growth was somewhat less successful from an investor's perspective. In 1897 the average return on investment in electrical utilities was only 4.02%, a much lower return than available from lower-risk railway bonds then being issued (Hausman & Neufeld, 2002), and a rate much lower than the stable returns utility investors would become accustomed to later in the twentieth century. There were many issues that this new industry struggled with, but there were two factors that made this marketplace such a difficult environment in which to be profitable: rapid technological change, and utilities' non-exclusive franchise.

Technological Change

Incumbents often enjoy protection from new entrants in a competitive market typified by high levels of fixed cost (Porter, 1979). However, that advantage can quickly evaporate in an environment of rapidly evolving technologies and declining unit costs, and that fixed cost investment that first formed a protective barrier can become an anchor,

https://doi.org/10.1515/9783110714036-008

tying companies to outmoded ways of doing business. Technological advances in the early years of electrical utilities caused capital equipment to be quickly made obsolete, and eroded those advantages traditionally held by an incumbent. Thomas Edison's DC-based central station grids, introduced only in 1882, were essentially made obsolete by Westinghouse's AC plants after only a little more than a decade. Despite having the first-mover's advantage of a virtual monopoly on central grid-supplied electricity for several years after the opening of Pearl Street, the eventual crush of technological competition with Westinghouse and Thomson-Houston would take a heavy toll on Edison's finances over the next decade (Wasik, 2006). By 1892 Edison had lost control of his electrical business to a consortium of financiers led by JP Morgan and, although he retained a 14% interest in the newly named firm, General Electric, his iconic name had disappeared from the new company and he was effectively removed from its management (Wasik, 2006). (However, Edison's name would live on in the many electrical distribution franchises already established across the United States.)

Even George Westinghouse's success did not guarantee financial viability, as in 1893 his company's short-term liabilities were found to exceed his assets by half a million dollars (Munson, 2005). *The New York Times* at that time reported that "financiers and stockholders now favor the appointment of a receiver for the Westinghouse Electric Company" (Munson, 2005, p. 35). Fortunately, Westinghouse was able to renegotiate his debts and save his company, saved in no small part by Nikola Tesla's offer to suspend the collection of royalties on his inventions, a gesture estimated by his biographer to have cost Tesla $17.5 million (Munson, 2005). The gesture seems particularly generous, considering that Tesla died in debt at the age of 86 (Jonnes, 2004).

Nonexclusive Franchise

A second contribution to the low profitability of the utility sector was the highly competitive market structure in which new electrical firms often operated. Electrical utilities typically commenced operations with the grant of a franchise from a city or municipality, providing the utility the right to dig up city streets, or string cables overhead, and operate the system for a period of years. In return, the utility undertook certain obligations, often including a commitment to maintain prices below designated price ceilings (McDermott, 2012). However, to encourage a competitive marketplace, these franchises were often not exclusive, and a city might have dozens of companies selling electricity within its territory (Demsetz, 1968). Often the franchises had overlapping territories, resulting in duplication of distribution assets. For example, Duluth, Minnesota, had five competing electric lighting companies before 1895, as did Scranton, Pennsylvania, in 1905 (Demsetz, 1968). New York City granted twenty-five nonexclusive franchises between 1882 and 1900 (Hausman & Neufeld, 2002), while Chicago similarly granted forty-five nonexclusive franchises between 1882 and 1905 (Swartwout, 1992).

The overlapping jurisdictions in which firms operated led not only to inefficiencies and intense competition, but also often resulted in an environment with multiple standards that made it impossible to achieve system-wide efficiencies. Even as AC became the most common operating standard, competing AC-based systems in the same city might each be operating at frequencies of 40, 60, 66, 125, or 133 Hz (Munson, 2005). Those cities that still ran legacy DC systems might deliver power at 100, 110, 220, or 600 volts (Munson, 2005). Neighbors on different sides of the street might find that the light bulbs or appliances purchased to operate on one side of the street could not operate on the electrical system across the street (Munson, 2005). In some neighborhoods the distribution cables of competing utilities would run down the same street, sometimes side-by-side with the cables of firms generating private electricity, such as streetcar companies. Furthermore, if one of the competing electric companies did go out of business, its deenergized electric plant and wires were often simply left in place (Bakke, 2016). In addition, the cables of the emerging telephone and telegraph companies often competed for the same locations above the sidewalk, and the space above many city streets was a congested mess (see Figure 6.1) (Bakke, 2016).

Figure 6.1: Overhead lines on Broadway, New York, late 1800s.
Source: Image courtesy of Museum of Innovation and Science, Schenectady, New York.

From a business model perspective, competitors often delivered different value offerings, generally with unique value chains. Today, we are accustomed to an electrical utility using common infrastructure to service all customers, from large factories to buildings needing a single light bulb. However, in the early days of the industry, legacy DC systems, newer AC systems, streetcar companies, or arc lighting systems,[1] all had their had their own generating plants and distribution systems, often delivering electricity at unique voltages and frequency to different customers. In addition, even as late as 1922, a quarter of the electricity in America was still being generated privately (Joscow, 1989). The opportunities for supply chain efficiency gains from shared infrastructure were being squandered, and although firms had largely agreed on AC technology as a standard for new distribution systems, it would take many years before the potential of true standardization would be attained. In these early decades of the industry, there started a growing realization that fragmented competition may not have been the most efficient method of organizing the business of electrical generation and distribution (Bakke, 2016), and perhaps the industry's dominant business model needed to be rethought. Once again, Sam Insull would be at the forefront of that deliberation.

Sam Insull (see Figure 6.2) had played a leading role in the previous business model transition of the 1890s, moving from a business model based on DC systems to one based on AC technologies. Now, decades later, he would lead the movement for another business model transition. Insull had risen from a position as a private secretary in Thomas Edison's London offices (Wasik, 2006), to become one of the lions of American industry. He had both wealth and power, enjoying a personal worth of more than $150 million by September 1929 (Wasik, 2006) (or more than $2.2 billion in 2020 dollars [US Bureau of Labor Statistics, 2020]) and ran utilities that served four million customers in thirty-two states, generating one eighth of the country's electricity (Taylor, 1962). His advice on the development of the industry was sought by foreign governments, including Canada and the United Kingdom (McDonald, 1957) and he was a regular visitor at the White House (Cudahy & Henderson, 2005). However, the industry that Insull served as a spokesman had struggled with persistent low profitability, and once again Insull would be a leading protagonist in the industry's pursuit of a new business model for the industry. Unlike previous transitions of the business model, which had featured new technologies and value offerings, Insull's proposal would

---

1 Arc lighting provided a very bright light widely used outdoors until the early 1900s, but poorly suited for most indoor uses. Robert Louis Stevenson particularly disliked the glare of arc lighting he had experienced at the Paris Opera, and declared that it "shines out nightly, horrible, unearthly, obnoxious to the human eye; a lamp for a nightmare! Such a light as this should shine only on murders and public crime, or along the corridors of lunatic asylums, a horror to heighten horror" (Stevenson, 1881, p. 255). One hundred and forty years before modern social media, Stevenson had already mastered the art of the negative review.

focus on government regulation to support a new business model adapted to operate as a regulated monopoly.

Figure 6.2: Samuel Insull, 1920, Age 60.
Source: Wikimedia Commons
https://commons.wikimedia.org/wiki/File:Samuel_Insull.jpg

The State of Early Regulation

Electrical utilities in the United States were only modestly constrained by regulation in the early years of the industry. Regulation was generally performed by cities or municipalities, and the regulation was most often exercised through control of franchise licenses (Knittel, 2006). However, this loose form of municipal regulation came with two notable problems: franchise risk and political influence (Hausman & Neufeld, 2002).

Franchise Risk

In the early years of the electrical utility industry in the United States, many cities and municipalities were legally restricted by state legislation from entering a contract (or awarding franchises) with a duration of greater than twenty years. The cost structure of an electrical utility, however, is dominated by investment in fixed assets, often with the lifespan of equipment measured in multiple decades. This mismatch between equipment life and franchise duration made raising capital difficult. Imagine the plight of a utility, with only ten years left in its franchise, trying to issue twenty-year bonds to invest in equipment with a twenty-year life. The risk of nonrenewal of the franchise would require a considerable risk premium from investors, and ultimately an increased cost to the consumer. In addition, as we discussed earlier in this chapter, the franchises were generally non-exclusive. For example, in 1880 Denver, Colorado, offered the grant of a franchise to "all comers" who wished to establish an electrical utility in their city (Hausman & Neufeld, 2002). Competitive abuses between utilities were not uncommon, and when consolidation did occur, it would often lead to the emergence of an effective monopoly in the market in which the remaining utility operated (Hausman & Neufeld, 2002). The situation would be neither attractive to early investors who had to bear the risk of a nonexclusive franchise, nor later to customers purchasing electricity from what was effectively an unregulated monopoly.

Insull's recognition of the value of a secure franchise may have been affected by his observation of the regulation of Chicago's streetcar system, a business that he would have been intimately familiar with as a supplier of electricity. During the 1890s, a fascinating entrepreneur by the name of Charles Yerkes consolidated a collection of small streetcar companies in Chicago into what was probably then the world's most efficient transit system (Cudahy & Henderson, 2005). However, each of the transit companies under Yerkes's umbrella operated with its own franchise agreement which, under law at that time, could not exceed twenty years. Every few years, the franchise agreement for one of the streetcar companies in Yerkes's system would expire, and Insull would observe Yerkes's exposure to recurring extortion from local politicians as he attempted to extend the franchise. The relatively short terms of the franchise agreements made it difficult to issue long-term debt for investment in expansion and renewal of the transit system. Yerkes proposed regulatory reform legislation that would create a more favorable environment in which to raise capital, by placing oversight of the transit system with a nonpartisan state-administered regulatory commission in exchange for a fifty-year franchise (Cudahy & Henderson, 2005). In the end, the legislation failed to pass into law, and by the early 1900s the transit companies were struggling to remain solvent. After a great deal of colorful scandal, Yerkes left Chicago to a more favorable investment climate in England where he became a founder of the company, "Underground Electric Railways Company of London Limited," which built many of the transit lines that operate in London to this day (Wasik, 2016).

Political Influence and Corruption

The second notable problem with early municipal regulation came from political influence and corruption. Many of the tools that are used today to mitigate political influence or corruption, such as financial transparency or professional regulatory management, were absent. Local politicians often carried out the administration of regulation themselves, leading to politicization of the process, and sometimes crossing into outright corruption (Troesken, 2006). As Sam Insull's biographer noted, utilities were easy targets for an ambitious politician:

> At the turn of the century, public utilities were regulated by municipal governments. Such regulation was governed largely by political concerns; shrewd politicians . . . recognized . . . that voters were often inclined to respond favorably to attacks on utilities.
>
> (McDonald, 1957, p. 117, as cited in Troesken, 2006, p. 270)

Utilities played the role of both victim and architect in the exercise of political influence and corruption during this period (Hausman & Neufeld, 2002). Sometimes, utilities were the victims of politicians who would use their regulatory authority to win favor over the electorate with promises of lower prices or enhanced services (Troesken, 2006). This type of political coercion of the electorate still exists with modern-day regulation, as evidenced by government-imposed price caps on consumer energy prices in Spain and Australia in the early 2000s, Illinois in 2016, or Alberta in 2017 (Brown, Eckert, & Eckert, 2017). In the early days of the industry, as today, these price controls could legitimately protect consumers from predatory monopolists, but often they were an easy political ploy to gain electoral favor. Whatever the political motivation, they would also produce distorted market signals, leading to inefficient consumption and investment decisions (Brown, Eckert, & Eckert, 2017).

Sometimes the utilities were the victim of outright political corruption. A notorious example was provided in Chicago, with a group of city politicians labeled the "Gray Wolves," which included men such as "Hinky Dink" Michael Kenna, and "Bathhouse" John Coughlin.[2] This group of politicians developed a scheme to grant a shell company, that they controlled, a long-term franchise to deliver service in the territory of an existing utility. This new company and its franchise would then be offered for sale at a substantial price to the existing utility (McDonald, 1957). The Gray Wolves had success with this scam in 1895, extracting a $7.3 million buyout for a shell company from an existing Chicago gas utility (Wasik, 2006). In 1897 the Gray Wolves tried this tactic again, granting a shell company that they controlled a fifty-year franchise for the delivery of electricity within the city of Chicago, with the intent of offering the company for sale to the existing electrical utility. This time, however, they were facing Chicago Edison's Samuel Insull, and it appears that Insull was prepared for this attempted extortion. Insull had managed to quietly secure exclusive licenses from every major American manufacturer of electrical equipment except Westinghouse, ensuring that Chicago Edison would have a near-exclusive right to install generation and distribution equipment in Chicago. Although the Gray Wolves were able to form the shell of an electrical company, Insull had impeded their ability to generate electricity without his cooperation (Wasik, 2006). Within a year Insull had purchased the shell company, Commonwealth Electric, and its franchise for $50 thousand, an incredible bargain for a fifty-year franchise, and eventually merged it with Chicago Edison to form Commonwealth Edison (ComEd) (Wasik, 2006).

---

2 Author's advice: When dealing with men named Hinky Dink and Bathhouse, watch your wallet.

In addition to playing the role of a victim, there were also many times when the utility was a full participant in influencing regulators or politicians. Sometimes the participation went to the level of corruption, as in San Francisco in 1906 where, in return for reneging on an election promise to lower the local utility's rates, fifteen of the sixteen city supervisors were found to have accepted kickbacks from Pacific Gas Light Company (Troesken, 2006). Sometimes, however, the utility engaged in political influence, without actually breaking laws then in effect. An example was set in 1926, when Mr. Frank L. Smith, who had been the chairman of the Illinois state body regulating electrical utilities, including Sam Insull's holdings, ran to represent Illinois in the US Senate. Although Mr. Smith won his election, the US Senate refused to allow him to take his seat when it became known that Sam Insull had supported Mr. Smith's Senate campaign with what was described as a $125,000 slush fund (Cudahy & Henderson, 2005). The ensuing political scandal would lead to Mr. Smith's eventual resignation of the Senate seat in 1928 without ever having occupied it, and Sam Insull was not found to have breached any laws. However, the public outrage from the incident led to greater calls for greater public scrutiny of utility holding companies, eventually contributing to the Public Utility Holding Company Act of 1935 (PUHCA) and the increased regulation of utilities (Cudahy & Henderson, 2005).

Today, the maturation of the regulatory process has greatly reduced the trade in influence between politicians and utilities. Sadly, however, improper influence still does sometimes still occur, as apparently evidenced by the 2020 criminal charges leveled against the speaker of the Ohio House of Representatives. The charges accuse the speaker of receiving the benefit of a $60 million dollar slush fund from one of the state's largest electrical companies, in return for pushing through a $1.3 billion bailout package for money-losing coal and nuclear generation plants (Gillis, 2020). Human temptation appears to be still with us.

A Proposal for Change

The electric utility sector in the early decades of the 1900s was an engine of economic growth. In the United States between 1900 and 1910, generation of electricity increased by 280%, from 4.5 million MWh to 17.2 million MWh, while generation capacity increased by 375% in that same period (Bradley, 1996). Average electricity prices fell by 41% in fifteen years, from 3.36 cents per kWh in 1902 to 2.89 cents in 1907, and 1.97 cents in 1917 (Bradley, 1996). Driven by declining costs and expanding grids, electricity was displacing kerosene and gas in lighting, and by the mid-1920s homes were installing new electrical appliances such as irons, toasters, and washing machines (Cudahy & Henderson, 2005). In manufacturing, as the grid expanded, electric motors replaced steam and water as the prime means of driving machines in the nation's factories (see Figure 6.3) (Devine, 1982).

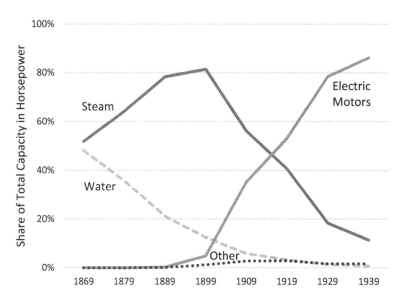

Figure 6.3: Sources of Mechanical Drive in US Manufacturing Establishments, 1869 to 1939. *Data Source:* Devine, 1982

As typical in any industry with a vertically integrated, capital-intensive value chain, this rapid growth would result in significant funding requirements, and between 1902 and 1917, the value of investment in plant and equipment in the electric utility industry grew at an average rate of 12% per year (Hausman & Neufeld, 2002). Electrical utilities were significantly more capital hungry than typical for other sectors of the economy, requiring a much greater investment to generate a dollar of revenue than other growing sectors such as steam railroads, telephone systems and street railways (see Table 6.1) (Hausman & Neufeld, 2002).

However, this cash hungry industry was not generating significant profits. In 1902, for example, the approximate 2,800 privately owned utilities in the United States had invested $483 million in plant and equipment in the prior twenty years but were generating revenues of only $79 million and profits of about $16 million (Hausman & Neufeld, 2002). Even with the rapid growth in sales, profit margins were not sufficient to fund the industry's capital requirements out of retained earnings (Hausman & Neufeld, 2002). To continue to invest in new, efficient plants and to continue growth of the grid to capture scale efficiencies, utilities would need access to equity or bond markets to fund capital requirements (Hausman & Neufeld, 2002). However, as we discussed previously, the environment in which the utility business model operated at that time was not terribly attractive to investors, with risk exacerbated by rapid technological change and by uncertain franchise durations. In addition, electrical utilities at the turn of the century tended to be local firms, without a national presence that would enable access to major capital markets for financing (Hausman & Neufeld, 2002).

Table 6.1: Ratio of Value of Capital to Value of Output (1929 US Dollars).

Year	Note	Electric Light and Power	Steam Railroad	Telephone	Street and Electric Railway	All Manufacturing	Chemicals	Agricultural Machinery	Motor Vehicles
1895		17.48	10.17	4.42	5.94				
1900		12.48	6.43	4.12	6.85				
1905	A	10.24	4.71	2.89	6.30	0.89	2.71	3.49	2.71
1910	A	10.47	4.35	2.54	5.77	0.97	2.13	3.33	2.02
1915	A	10.26	4.34	2.26	5.12	1.01	2.30	3.59	1.21
1920	A,B	4.51	3.17	1.58	4.04	1.02	1.84	1.72	0.88

Notes: Data refers to one year earlier for the following categories: All Manufacturing, Chemicals, Agricultural Machinery, and Motor Vehicles in the years 1905 to 1920. By 1920, most US states had adopted state-wide regulation of electric utilities.
Source: Hausman and Neufeld (2002), p. 1053, which is citing sources for utilities and railways data from Ulmer (1960), pp. 256–257, 320, 374–375, 405–406, 472–473, 476, 482, 486. and sources for manufacturing data from Creamer, Dobrovolsky, and Borenstein (1960), pp. 265–267.

Historians have identified different ways in which utilities attempted to address the need for financing in this period. One method of securing financing during this period was provided by the manufacturers of electrical equipment, who would offer to receive stock and bonds in the utilities as payment for equipment (Hausman & Neufeld, 2002). In some cases, the manufacturer would then bundle securities from several utilities into an investment trust, made available to other investors. This eventually evolved into a holding company structure, which might then issue its own equity and bonds to further finance the acquisition of operating utilities companies (Hausman & Neufeld, 2002). Holding companies would also raise capital in the operating utilities using preferred or non-voting shares, allowing investors in the voting shares of the holding company to maintain control of the total organizations with a relatively small investment. For example, Sam Insull used a holding company structure to control $500 million in utility assets at the end of the 1920s with an equity investment of only $30 million (Lambert, 2015).

The use of holding companies did not necessarily alter the fundamental business model of the utilities operating under the umbrella of the holding company, but it did provide capital, and as a financing structure it came to dominate the industry by the end of the 1920s. The operating utility companies under the umbrella of the holding company were typically not integrated, so did not gain integration efficiencies (Hausman & Neufeld, 2002). However, since the holding company was seen to bring expertise and efficiency in finance, management, and engineering, its securities were seen to be attractive by the investing public (Hausman & Neufeld, 2002). For example, the holding company controlled by Sam Insull in the 1920s, Middle West Utilities, attracted retail investors with the slogan "If the light shines, you know your money is safe" (Cudahy & Henderson, 2005). However, many of these holding companies collapsed in the aftermath of the Great Crash of 1929, and the structure was effectively banned by the Public Utility Holding Company Act (PUHCA) in 1935. Although the development of the holding company structure was successful for a short period in arranging access to capital for utilities, it could be argued that it never fundamentally changed the dominant business model of the operating utility. That fundamental change was still to come and would again be led by the same person so closely identified with the last major change: Sam Insull.

Sam Insull Seeks a Monopoly

Through most of the first several decades following the implementation of the Pearl Street grid, the industry remained very lightly regulated, with multiple electric companies often operating and competing in a single territory. However, among privately owned electrical utilities, Samuel Insull, the head of the Chicago Edison Company, became the first leader of a major utility to advocate for state regulation of the electric industry (Wasik, 2006). Even as early as 1897, Insull declared in his first major speech to the industry's leading organization, the National Electric Light Association (NELA), "Our business is a natural monopoly, it must be of necessity regulated by some form of governmental authority" (Wasik, 2006, p. 79).

Insull's thinking further developed over the following year, and in June of 1898 he went back to NELA as its new president, and laid out his vision for a new industry business model in a controversial speech that, in retrospect, would be viewed as a seminal event in the industry's development (Cudahy & Henderson, 2005). In this 1898 speech Insull argued that competition was leading to higher costs of financing the industry's electrical grid:

> the interest cost on their product (which is by far the most important part of their cost) is rendered abnormally high, owing partly to duplication of investment and partly to the high price paid for money borrowed during the period of competition.   (Insull, 1898, para. 13)

Insull argued that this competition ultimately led to increased prices for the consumer, noting that:

> if the item of interest be necessarily augmented, it must be reflected in the price paid by public and private users.   (Insull, 1898, para. 14)

Insull further argued that consumers would benefit from lower costs and better service under a full monopoly:

> the best service at the lowest possible price can only be obtained by exclusive control of a given territory being placed in the hands of one undertaking.. . . the more certain this protection is made, the lower the rate of interest and the lower the total cost of operation will be, and, consequently, the lower the price of the service to public and private users.
> (Insull, 1898, paragraphs 15–17)

However, Insull also recognized that elimination of competition would require greater public regulation. Insull argued:

> In order to protect the public, exclusive franchises should be coupled with the conditions of public control, requiring all charges for services fixed by public bodies to be based on cost plus a reasonable profit. It will be found that this cost will be reduced in direct proportion to the protection afforded the industry.   (Insull, 1898, para. 16)

So, Sam Insull, one of the great deal makers of his age, proposed a business model that would represent a radical new deal between the utility and the ratepayer. In an arrangement between utility, ratepayer and regulator that would evolve into a relationship known as the "regulatory compact," the utility would receive a grant of monopoly status in exchange for ratepayers receiving the benefit of lower costs and universal service. The reasoning that Insull laid out in 1898 was visionary. However, the vision was not immediately shared by his industry peers who, at that time, favored the status quo of local regulation and competition over the new risks posed by much more rigid, systemic regulation (Bradley, 2011). Nevertheless, over the following decades, Insull's peers and counterparts in government did eventually come to support the logic of regulation. As we will discuss later in the chapter, by the 1920s, a consensus of support started to emerge in the utility industry and with legislators supporting monopoly regulation (Wasik, 2006). By the mid-1930s, following the tumultuous events of the great stock market crash, the regulatory framework that Insull had imagined decades earlier would be firmly in place.

Benefits of the New Business Model: From the Utility Perspective

Slowly, and over many years following Insull's 1898 speech, consensus among the utility leadership evolved to support the idea of regulation. However, the utilities' motives were not based entirely on altruism, as a regulated monopoly structure could offer significant sources of benefit to the industry. First, regulation and provision of an exclusive operating territory acted as a means of establishing barriers to entry by potential competitors. Second, state-wide or province-wide regulation would be easier for utilities to manage than an assortment of regulatory frameworks that existed in different city, town, and rural service areas. Third, state-wide bodies run by professional management would be less susceptible to adverse influence than local politicians to gain political favor. Fourth, privately owned utilities saw regulation as a means to discourage expansion of public ownership of utilities. The number of utilities in the US owned by cities or municipalities in this period was significant, although they tended to be smaller than their counterparts in the private sector. In 1902, 22.5% of the utilities in the country were municipal, although they accounted for only 7.8% of the electricity generated (Jacobson, Klepper, & Tarr, 1985). By 1927, the share had grown to 50.7%, although now accounting for 4.5% of electricity generated (Jacobsen et al., 1985). However, in many other countries, governments had taken a much more active role in establishing a publicly owned electric sector. In Canada, Ontario Hydro had been established in 1910 as a publicly owned utility with the mission of establishing economic development through electrification, followed by similar initiatives in other Canadian provinces (Nelles, 2003). The public ownership of a large public enterprise like Ontario Hydro, with its low customer rates and engineering prowess, stood out as an "object of anathema" to the American industry trade group, the NELA (Nelles, 2003, p. 121). The periodic comparison of electricity rates between public and privately owned utilities would sometimes reach the attention of the public, and electricity rates charged by investor-owned utilities were often found to be significantly higher than those of their municipally owned counterparts. For example, a late 1920s investigation into the rates of investor-owned Commonwealth Edison in Chicago found electricity rates to be eight times those of a neighboring municipally owned utility (Wasik, 2006).

Public sentiment is a critical and probably undervalued asset for a utility, and a utility that loses it can become a target for politicians. The prospect of a government revoking the franchise of a poorly performing utility to be replaced by a publicly-owned entity might seem remote, but it did happen, and still happens today, as seen in the 2019 threat by the Governor of New York to revoke the licence of a gas utility serving 1.8 million New York customers from $6.3 billion rate base (Balaraman, 2019). In the 1920s, many leaders of electrical utilities could see that public sentiment toward privately held electric providers was becoming increasingly negative, and the threat of forced conversion to municipal ownership was real (Wasik, 2006). Submitting to greater regulatory oversight was one way for the utilities to reduce that threat, and it would not be the first-time regulation had been used to isolate an industry from public scrutiny. In 1892, responding to a railway CEO who wished to abolish the federal commission governing the operation of railways, the United States Attorney General in the cabinet of Grover Cleveland, Richard Olney, noted the value of regulation to railway interests.

> The [Regulatory] Commission . . . can be made, of great use to the railroads. It satisfies the public clamor for a government supervision . . ., at the same time that that supervision is almost entirely nominal. Further, the older such a Commission gets to be, the more it will be found to take the business and railroad view of things. It thus becomes a barrier between the railroad corporations and the people, and a sort of protection against hasty and crude legislation hostile to railroad interests. The part of wisdom is not to destroy the Commission but to utilize it. (Bernstein, 1955)

As a further benefit to utility shareholders, and one of Insull's prime arguments for the proposed new business model, the revenue stability that would come with regulation would greatly reduce the risk of investing in utility debt and equity (Hausman & Neufeld, 2002). Along with the demand for electricity, the demand for financing was growing. In the early days of the Edison's central station grid, the amount of financing required to construct a central station grid that served a few square miles was relatively modest. However, with transition to larger AC systems, and the continued growth of plant size to capture scale efficiencies, generation plants and grids were much larger, and utilities needed access to bond markets to meet growing capital costs. The uncertainty inherent in the limited duration of the franchise in a capital-intensive industry, plus the pressure on profits of competing in a non-exclusive franchise, greatly increased borrowing costs. This, in turn, increased costs to consumers and reduced the returns that could be paid to shareholders (Hausman & Neufeld, 2002).

Benefits of the New Business Model: From the Legislator's Perspective

From the legislator's perspective, there were several factors that supported broader and more active regulation of the industry. The first was an emerging idea that wide-spread distribution of electricity was in the public's interest and should be encouraged. This concept of government actively facilitating the wider distribution and consumption of any product represented a somewhat dramatic shift in traditional American public policy. Traditional perspectives on government's role in the economy was one of laissez-faire, and this had been particularly reflected in and reinforced by successive decisions of the US Supreme Court up to that time (Tomain, 1997). The idea that government should have a role in promoting the consumption of a product was novel, and would later become central to the Roosevelt's New Deal regulatory programs of the 1930s (Tomain, 1997). The idea of cheap electricity as a tool to achieve social betterment joined public transit, public health, and improved public services as worthy government objectives (Nelles, 2003).

The second factor supporting broader regulation of the industry was rooted in the idea that the market for production and distribution of electricity was a "natural monopoly" and should be regulated as such. The concept of a "natural monopoly" had been described by John Stuart Mill in the 1848 as those monopolies "which are created by circumstances, and not by law" (Mosca, 2008, p. 323). Economic theory of the time predicted that a single firm operating in a market that is a "natural monopoly" could produce more efficiently and at lower cost than if the market was populated with multiple firms (Mosca, 2008). Many politicians surely could see the evidence of that argument, just by walking down a street which carried the redundant distribution networks of two different utilities, and the wasteful duplication of costs from having more than one competitor in what appeared to be a natural monopoly. However, economic theory also predicted that the strongest firm in a natural monopoly, if not regulated, would eventually come to dominate its market, and exercise monopoly pricing power to the detriment of consumers (Tomain, 1997). Accordingly, by prevailing political thought at that time, firms operating as natural monopolies should be regulated to deliver services to customers at the lowest possible cost, while being fairly compensated. Governments had already stepped in to regulate railways, gas distribution and shipping, and regulation of the distribution of electricity would be a natural extension of that practice.

Finally, politicians and the electorate appear to have recognized the temptation of political influence, and the benefits that a non-partisan regulatory body could bring in removing politicization from the regulatory process. The promise of these state-wide regulatory bodies was reflected in this state-level Republican party platform of the time:

> We advocate a just, impartial, and unprejudiced control of public service corporations and public utilities generally in this state through incorruptible, enlightened, and non-partisan agencies; and we condemn any exemption from such supervision and control, and any other special favors to any particular enterprise or corporation.   (Bradley, 1996, p. 64)

The Regulatory "New Deal"

The boundaries of early regulation of electrical utilities most often reflected the natural boundaries of the underlying distribution technology. Initially, utilities were restricted in the distance that they could distribute electricity, and regulatory authority was exercised by the city or municipality in which the utility resided. However, as utilities started to extend outside of city limits it became natural that larger political jurisdictions, such as states and provinces, would get involved. This started in the United States in 1907 when Wisconsin, Georgia, and New York expanded the regulatory scope of their railway commissions to include electricity (Knittel, 2006). Other states' regulatory bodies followed, and by the end of 1922, there was state level regulation of electrical utilities in 37 states[3] (Knittel, 2006). The initial mandate of these new state-level regulators was like those of their city or municipal predecessors, and was focused on the establishment of rates, standards and controlling the entry of new competitors (Knittel, 2006). However, unlike municipal-level regulation which had often been administered as part of the job of local politicians, these state-level regulatory bodies provided full-time professional management that would develop the new regulatory framework that would define a new business model for electric utilities.

Although Insull's seminal speech advocating a new industry business model had been given in 1898, and state regulatory bodies started to be established in 1907, it was not until the mid-1920s that a consensus started to form among policy makers and industry leaders that a significant new regulatory framework was required (Tomain, 1997). Legislators recognized that the most cost-efficient delivery of a very desirable public good, electricity, would be best accomplished through a monopoly service provider (Tomain, 1997). Utility management and shareholders saw the benefits of the elimination of competition, reduced borrowing costs and stable rates of return. With public and private interests aligned, utilities and governments started to work toward a relationship that would come to be known as "the regulatory compact," a relationship that would form the foundation of traditional utility regulation to this day. This new relationship was based on a few fundamental ideas that would transform the dominant utility business model. First, the utility, would be awarded a *monopoly franchise* to distribute electricity within a certain jurisdiction. Second, the utility would take on an *obligation to serve* all customers within that service jurisdiction who desire to receive their service, while accepting regulatory oversight of its activities (Tomain, 1997). Third, the regulator would allow the utility to sell electricity at a price that allowed recovery of its *cost of service*, which includes a fair *rate of return* on invested capital that is *prudently* invested in assets that are *used and useful* (Tomain, 1997).

---

3 Transmission of electricity in the United States across state lines in these early days was limited, and even with the creation of the Federal Power Commission in 1920, federal involvement in the regulation of electricity remained low for many years (Knittel, 2006).

These simple sentences contain concepts that have taken thousands of pages of jurisprudence and decades of regulatory deliberations to refine. In fact, these regulatory concepts have been key in forming the dominant utility business model that we have today and, as discussed later, often form impediments to the evolution of today's utility business model to meet the changing environment of the twenty-first century. Accordingly, let us briefly examine these concepts in a little more detail to get a better understanding of their meaning, history, and impact on the dominant utility business model: *monopoly franchise, obligation to serve, prudency of investment, used and useful, rate of return,* and *cost of service.*

Monopoly Franchise

Prior to the 1920s, municipalities had often awarded electrical utilities with nonexclusive franchises, with the objective of spurring competition and lower electricity prices within their jurisdiction. However, the prospect of available scale economies combined with the reality of low profitability, provided a businesses environment ripe for consolidation. In fact, rather than encourage competition, the consolidation of competition combined with light regulation, resulted in the distribution of electricity in some markets being serviced by a de facto unregulated monopoly. The award of a monopoly franchise by a government sometimes formalized what was already effectively a reality.

The award of a monopoly franchise would not be a wholly new approach by governments to natural monopolies, and had been developed decades earlier with gas distribution companies. In 1812, a German engineer and inventor, Frederich Winsor, had been awarded an exclusive twenty-one-year, royal patent to distribute gas in London, and by 1815 had built a distribution network of 25 miles (about 40 km) of gas mains under the streets of the city (Nelles, 2014). Over the course of the nineteenth century, the establishment of similar regional monopolies became common industrial policy throughout Britain and then the world. By 1827, New York's gas monopoly was lighting Broadway with hundreds of gas lights, and in 1828 the Champs Elysees was similarly illuminated (Nelles, 2014). European and US industrial and regulatory experience was exported around the world with the construction of a French designed gaslight company in Saint Petersburg in 1835, the Montreal Gas Light Company in 1846, the British-owned Hong Kong and China Gas Company in 1862 and the Sao Paulo Gas Company in 1869 (Nelles, 2014). Application of the monopoly franchise to the electricity industry was a natural extension of a long-established approach.

How did the award of monopoly franchise impact the typical utility business model? One feature of a utility's business model that would be radically altered was its *value capture* activity. The days of using price to establish a position in a competitive marketplace were over. Instead, the utility would determine prices through a regulatory process. Another impact on the business model was the limited *customer identification* that arose from placing geographic boundaries around the utility's service territory. However, the revenue stability that came with that monopoly franchise would open capital markets to utilities and enable investment in expansion of the grid within those geographic boundaries. In addition, since utilities no longer competed directly against each other, the *value chain* could be made more efficient, as duplicate assets could be eliminated, greater scale economies would be accessible, and cooperation between utilities in the establishment of common standards and practices would become much more likely.

Obligation to Serve

In return for receiving an exclusive right to sell electricity in a certain geographic area, a utility is obliged under the "regulatory compact" to provide electrical service to those customers in their service area who request it, at a rate that is regulated (Joskow, 1986). The obligation was well described in a foundational Canadian case, *Chastain v. British Columbia Hydro and Power Authority* (1972):

> The obligation of a public utility or other body having a practical monopoly on the supply of a particular commodity or service of fundamental importance to the public has long been clear. It is to supply its product to all who seek it for a reasonable price and without unreasonable discrimination between those who are similarly situated or who fall into one class of consumers.

Regulators and courts have placed practical limitations on this obligation to serve, as will be discovered by utility customers who fail to pay their bills on a timely basis. In addition, the regulator will generally limit the extent to which a utility must expand its system to service remote customers, and it may require a new customer connecting to the system to pay all costs incurred to expand the system to the customer's premise. Nevertheless, when imposed in the 1920s and 1930s, this obligation to serve affected many aspects of the firm's business model. First, it impacted utilities' *customer identification*, as firms could no longer choose to maximize profitability by targeting certain market segments as customers. Under the new business model, since utilities were restricted in their ability to refuse service to higher cost customers, the rates paid by lower cost customers would necessarily increase to maintain the utility's rate of return (Smith, 1996).

Second, the obligation to serve also impacted utilities' *value capture* by eliminating the ability to use price discrimination to attract profitable customer groups. Modern economics tells us that a profit-maximizing company in a competitive market should price above marginal cost to the maximum extent allowed by the customer's price sensitivity, or in other words, as determined by the customer's elasticity of demand (Hausman & Neufeld, 1989). In an unregulated, competitive marketplace, utilities had been able to use market segmentation and price discrimination to extract higher margins from certain groups of customers by charging higher prices than justified by the cost of service. Under the new regulatory compact, although a utility was required to set separate rate classes to reflect the costs of serving different groups of customers (e.g., different rates for residential, commercial, and industrial customers), its obligation to serve all customers effectively restricted its ability to use price discrimination to maximize profits (Smith, 1996). It also restricted for many years the ability of the utility to use pricing to attract consumption in off-peak periods, when the supply of electricity carries a much lower marginal cost (Hausman & Neufeld, 1989). Indeed, Sam Insull had achieved remarkable growth in the electricity sold by Chicago Edison in the 1890s by offering discounted, introductory contracts to large customers to induce them onto the grid (Lambert, 2015). However, under the new regulatory compact, it was not until the early 1980s that utilities in North America would widely implement load-control rates and time-of-use rates to once again incent ratepayers to shift consumption to off-peak periods, rediscovering some of the pricing tools that Sam Insull had used in Chicago in the 1890s (Hausman & Neufeld, 1989).

Finally, the obligation to serve impacted the *value offering* and *value chain* of the dominant utility business model, as it imposed requirements to meet reliability and quality of service standards that would be common across a customer group. Under the new regulatory compact, utilities and regulators focused for decades on the supply side of service delivery, ensuring a value chain that delivered reliable, standardized electrical service at the lowest cost (Smith, 1996). The product was also standardized so that, except for very large industrial users, the requirement to offer an undifferentiated product largely limited utility customers' ability to choose between price and a product attribute, such as reliability. Generally, customers were required to make their own arrangements to change the attributes of electrical service (Smith, 1996). For example, utilities have long known that different customers will place a different value on avoidance of electrical service interruptions. A local butcher shop, with electrically powered coolers and walk-in freezers, would be willing to pay more for reliability of electrical service than the clothing store next door. However, since they are unable to pay a higher price to the utility for a differentiated service with enhanced reliability, customers have been required to rely on other means to attain greater reliability (e.g., the butcher shop may install its own back-up generation). However, as we will discuss in later chapters, utilities in recent years have addressed growing customer demand for a differentiated product by offering a broader selection of ancillary services to supplement their standard electrical service.

Tests of "Prudence" and "Used and Useful"

A firm's management, operating in a competitive, unregulated market is constrained by its competitors in the actions it can take. Imprudent investments or operating decisions in an unregulated environment will result in reactions from competitors with financial consequence, and a loss of shareholder value. However, in a regulated monopoly market, the presence of competitive market forces is replaced by the regulatory test of "prudence" to ensure that management operates effectively and efficiently (McDermott, 2012). The legal concept of prudence is found in many aspects of common law, and as it relates to regulated utilities, there are subtle differences of interpretation between regulatory jurisdictions. Nevertheless, the basic interpretation of prudence is generally agreed across jurisdictions, and was given a useful definition in this regulatory proceeding:

> In summary, a utility will be found prudent if it exercises good judgment and makes decisions which are reasonable at the time they are made, based on information the owner of the utility knew or ought to have known at the time the decision was made. In making decisions, a utility must take into account the best interests of its customers, while still being entitled to a fair return. (Alberta Utilities Commission, 2001, p. 10)

In principle, the prudence of an investment must be judged based on the prevailing knowledge at the time the decision was made, and not based on a retroactive view of events. If an expenditure of the utility is judged not to be prudent, then the regulator may disallow the utility from recovering those costs, effectively placing the burden of the cost on the utility shareholders (Joskow & Schmalensee, 1986). The impact on utility shareholders of a failed prudency review can be significant. During the 1970s and 1980s, a sudden slowdown in demand growth combined with large cost overruns in coal and nuclear generation projects, resulted in over fifty major prudence reviews of electric utility project costs by US state regulators, with the disallowance of $14 billion in construction costs (McDermott, 2012). These costs were carried by utility shareholders instead of ratepayers.

A concept closely related to the prudence test is that assets included in the rate base should be "used and useful." The test is intended to ensure that ratepayers should only pay for assets that are currently being used by the utility to support delivery of services (Lesser, 2002). Like the prudence test, the failure of a utility to meet this test will also cause the costs of the financial investment to be borne by the utility's shareholders. However, unlike the prudence test, the "used and useful" test may cause regulators to disallow investments from the rate base that have become uneconomic due to changing circumstances, whether the investment was prudent when originally incurred, or not (Lesser, 2002). One well known example involved a group of Vermont utilities that entered a thirty-year contract for power from Hydro Quebec, commencing in 1990 (Lesser, 2002). When deregulation of natural gas in 1994 caused the cost of competing electrical supplies to fall, the regulator ruled that the contract with Hydro Quebec was no longer economically used and useful. As a result, the Vermont utilities were unable to recover the full cost of their contract with Hydro Quebec. This regulatory decision was highly controversial (Lesser, 2002), since to some observers it seemed to strain the assumption embedded in the regulatory compact of fair and equitable treatment of both ratepayers and utility shareholders, and that the assessment of a utility's investment decisions should not be based on a retroactive reading of circumstances. Regulators need to tread carefully in the application of the "prudence" test and the "used and useful" test in sharing risks between the two parties, or investors will be reluctant to invest without increased returns to compensate for increased risks (Lesser, 2002).

From a business model perspective, the test that an expenditure be prudent and the test that an asset be used and useful, form part of a utility's *value offering* to the rate payer. These tests provide the rate payer with a degree of assurance and protection from an undesirable cost-plus relationship at the hands of a monopoly service provider (Joskow & Schmalensee, 1986). Under cost-of-service regulation (which we discuss later), there is a natural temptation for a monopoly service provider to make unnecessary expenditures that would tend to increase customer costs, while increasing returns to the utility's shareholders. The regulator's punitive threat of disallowing the recovery of an investment's cost from the utility's customers acts as a counterbalance to this temptation, and motivates the utility to make sound decisions. However, as the author has discovered in his own research, and as discussed later, the threat of disallowance can also lead the utility to make decisions that support old business models and can discourage investments required to transition to new business models. In an environment of changing technology, whether it was the investment in nuclear power in the 1980s, or distributed energy resources in the 2020s, the presence of these tests can hinder restructuring of the industry (Lesser, 2002). A heavy-handed regulatory approach in applying the tests to the detriment of utility shareholders, may increase the reluctance of utilities to leave established technologies and to test innovative business models.

Cost of Service Regulation and Rate of Return

Even though states made first steps in establishing regulatory bodies in 1907, regulation of electricity distribution as we understand it today did not suddenly emerge fully formed with the creation of these first regulatory commissions. One of the key building blocks was the regulatory statute enacted in 1907 in the state of Wisconsin, drafted by John Commons, an economist at the University of Wisconsin. The legislation was foundational and widely emulated, encompassing factors such as monopoly franchise, obligation to serve, and perhaps most importantly, the requirement that utility rates be "reasonable and just" (Boyd, 2018, p. 756). As described by the Wisconsin Supreme Court in this 1923 judgement, this requirement would require a balance between the interests of the ratepayer and shareholder:

> It is the duty of the Commission to prevent unreasonable exactions by the utility on the one hand, and also to protect the rights of investors from confiscation by imposition of rates which are too low on the other. The rate should be in the language of the statute "just and reasonable."

In other words, not so low as to approach the line of confiscation nor so high as to be unjust and oppressive.

> (Waukesha Gas & Elec. Co. v. R.R. Comm'n of Wis., 194 N.W. 846, 849 (Wis. 1923), as cited in Boyd, 2018, p. 756)

Although the states' legislation provided early regulatory bodies with the power to regulate and to establish "just and reasonable" rates, the legislation was generally silent on how those rates were to be achieved (Covaleski, Dirsmith, & Samuel, 1995). It took many years for regulators and courts to develop a structure for setting rates while balancing the interests of ratepayers and shareholders. As regulatory mechanisms matured, the rate setting process extended beyond mere price caps to include the ability to set rates in a reasoned, nonarbitrary fashion.

The value capture element of the new business model that emerged from the regulatory compact would eventually be largely defined by Cost of Service Regulation (COSR), a framework that has implemented the financial bargain between utilities and ratepayers. Under COSR, the regulator works to achieve a balance, to ensure that ratepayers are protected from the utility's monopoly pricing power, while the utility can recover its prudently incurred costs plus a reasonable return on its shareholders' invested capital (McDermott, 2012). In simple terms, the revenue allowed to a utility under COSR is calculated as equal to its operating costs, depreciation, and taxes plus the allowed rate of return on the utility's rate base (McDermott, 2012). This can be represented by the following equation (McDermott, 2012):

$$RR = TC = (OC + D + T) + ((RB - D) \times ROR)$$

where,

RR = Annual Revenue Requirement (that determines the unit cost of electricity to the customer when divided by the annual quantity of electricity sold)

TC = Total annual cost of the utility's regulated operations
(OC + D + T) = Annual operating cost, depreciation cost, and taxes
(RB − D) = Rate Base, net of accumulated depreciation
ROR = Approved annual rate of return for the utility shareholders' investment

Although the concepts embedded in the equation seem simple and straight forward, their actual implementation has taken decades for regulators and courts to sort out, and the debate continues still. Initially, there were many fundamental impediments to implementing the COSR framework, such as a lack of financial transparency and an absence of standardized reporting. Even the use of historical cost as the appropriate basis for valuing the utility's capital investment would not be settled until the 1940s, almost four decades after Wisconsin's legislation established the requirement that regulated utility rates be "reasonable and just."

One of the early barriers to effective regulation was the lack of accounting disclosure provided by utilities in the early days of the industry. For example, the balance sheet presented by Commonwealth Edison in 1915 was twelve lines long, accompanied only with a two-page discussion of the company's financial state (Wasik, 2006). Even with these scanty disclosure standards, firms were sometimes reluctant to disclose any financial statements at all. Electric Bond and Share Company, a utility holding company formed by General Electric, only disclosed its financials after being taken to court in 1928 (Wasik, 2006). Insull similarly resisted fully opening his books until 1930 (Wasik, 2006). Although the grant of monopoly powers came with greater requirements for regulatory scrutiny of expenditures, the industry habitually resisted attempts to justify their rates to the public by the way of financial disclosure.

Related to this lack of accounting disclosure, was the absence of a regulatory tool we take for granted today: a uniform set of accounts (Covaleski et al., 1995). Without a uniform set of accounts, it was easier for the utility to hide information, and more difficult for the regulator to ask the right questions about the utility's financial disclosures. Comparison between utilities, or between years for the same utility was difficult. Establishment of a common set of accounts became a priority for the newly formed state-level regulatory commissions (Bradley, 1996). Eventually, with a uniform set of accounts and transparent financial disclosures, the private information of the utility became more public, and set the framework for discourse and investigation by the regulator (Covaleski et al., 1995).

Another of the early problems in establishing the COSR framework was calculating the value of the rate base on which the return should be based. In general, the rate base consists of the value of property used by the utility in providing service. Modern regulation recognizes this value as the depreciated cost of the investment that a public utility has made in regulated assets used to deliver service to the customer. It is on this depreciated cost that a public utility is permitted to earn a specified rate of return, in accordance with rules set by a regulatory agency (Ghadessi & Zafar, 2017). However, in the early years of the establishment of state regulatory bodies, utilities had been allowed to use either depreciated cost or market valuation of their investment as the basis on which "rate of return" should be calculated (Covalesk et al., 1995). This resulted in decades of debate as regulators and utilities grappled with the definition of market valuation (McDermott, 2012). The use of market valuation was not easily verifiable, and the valuation could lead to a circular logic: if the rate was dependent on an agreed return on market value, then surely the market value of the asset depended on the revenue that it could generate, which was itself dependent upon the rate that would be determined (Covaleski et al., 1995).

A crucial turning-point in American regulation occurred in 1944, with the US Supreme Court decision in *Federal Power Commission v. Hope Natural Gas*, which supported the use by regulators of historic, depreciated cost as an appropriate basis for calculating the utility's capital investment (McDermott, 2012). James Bonbright, possibly the most influential American regulatory economist of his generation, described the decision as "one of the most important pronouncements in the history of American law" (Boyd, 2018). This decision settled the debate: regulators could use historic, depreciated cost valuation in the calculation of rate base.

A further factor impeding the determination of electricity rates that were "fair and just" was the use of holding companies to enable the abuse of financial disclosures. Holding company structures in the 1920s were widely used to mask true asset costs and transfer prices, or artificially inflated or deflated prices in transfers between subsidiaries (Tomain, 1997). The holding companies became increasingly dominant in the industry, and by the end of the 1920s, ten holding companies controlled 75% of the production of electricity in the United States (Bakke, 2016), the three largest being Electric Bond and Share (owned by General Electric), United Corporation (controlled by JP Morgan), and the Insull Group (Bradley, 1996). However, the stock market crash of 1929 brought to light shareholder abuses and illegal financial manipulations at many publicly traded utility holding companies (Tomain, 1997). The reaction of the public to the financial downfall of publicly owned utilities in 1929 was taken up by politicians and set the stage for closer public oversight and regulation (Tomain, 1997). By 1935, the United States had enacted the Public Utility Holding Company Act (PUHCA), greatly strengthening regulatory oversight of the nation's utilities. The Act placed restrictions on the former "holding company" structure that had previously shrouded financial transactions and structures (Cudahy & Henderson, 2005), and limited utility holding companies to a single integrated public utility. It also forced utilities to follow more strict financial reporting rules established by the Securities and Exchange Commission (Cudahy & Henderson, 2005). As noted by Tuttle et al. (2016, p. 6), "the PUHCA[4] established the framework for the traditional electric utility network." Other countries, including Canada, soon developed regulatory bodies to model US regulatory frameworks, and rolled out similar implementations after studying US experience (Melody, 2002).

Despite the hurdles in their path, regulatory structures matured and became more effective. The regulatory process was dominated by regulatory professionals rather than local politicians more susceptible to politicization, and sometimes corruption, of the regulatory process (Troesken, 2006). The "regulatory compact" developed by these regulators and utilities formed a framework for a new business model (see Table 6.2) that would bend in certain areas over the years but would largely remain intact into the next century.

---

4 The Public Utility Holding Company Act of 1935 remained active for over seventy years, until its repeal on 8 February 2006, when it was replaced with a package of new regulations enacted under the Energy Policy Act of 2005, signed into law by President George W. Bush (Thakar, 2008).

Table 6.2: Key Influences of the "Regulatory Compact" on the Utility Business Model.

Element of the "Regulatory Compact"	Business Model Element Impacted			
	Customer Identification	Value Offering	Value Chain	Value Capture
Award of "Monopoly Franchise"	Customers identified by jurisdictional boundary.		Made more efficient, as duplicate assets eliminated, greater scale economies accessible, and common standards and practices implemented. Access to equity and debt finance enabled investment in value chain.	Ensures revenue stability. Price differentiation no longer an available tool to increase profitability. Limits profits by placing boundaries around customer identification and geographic expansion.
Utility assumes an "Obligation to Serve"	Obligation is to supply its product to all who seek it for a reasonable price and without unreasonable discrimination. No longer able to maximize profitability by targeting certain market segments as customers.	Requirement to offer an undifferentiated product. Most customers do not have an ability to choose between price and a product attribute such as reliability.	Requirements to meet reliability and quality of service standards that would be common across a customer group.	Eliminates the ability to use market segmentation to select higher margin customers. For many years restricted the ability of the utility to use pricing to attract consumption in off-peak periods.
Investments must demonstrate "Prudence" and that they are "Used and Useful"		Provides the rate payer with assurance and protection from an undesirable relationship with a monopoly service provider.	The threat of disallowance can also lead to decisions that support old business models and can discourage investments to transition to new business models.	

Table 6.2 (continued)

Element of the "Regulatory Compact"	Business Model Element Impacted			
	Customer Identification	Value Offering	Value Chain	Value Capture
Compensation through "Cost of Service Regulation" ("COSR")				Forms the core basis for determination of utility revenue and profit. COSR sometimes modified through tools such as performance-based regulation (PBR).

A New Dominant Model – Centralized Production, Regulated Distribution

Electrical utilities in the early part of the twentieth century were not only one of the fastest growing sectors of the economy, but electrification was also driving growth throughout the economy. Between 1910 and 1930, the share of factories using electric power grew from 25% to 75%, contributing to a tremendous boost in manufacturing, with an annual growth rate of 5.4% in manufacturing output between 1919 and 1929 (Gordon, 2004). And it was not just factories that were electrifying. Between 1912 and 1929, the share of American homes with electric service increased from 16% to 68%, (Gordon, 2004) introducing not only electric lighting into the home, but also electrical appliances such as such as vacuums, irons, iceboxes, and radios (Gordon, 2004).

With this growth in demand for electricity, by 1922, forty years after Thomas Edison had launched the first public electric utility on Pearl Street in Manhattan, the United States was home to 3,744 privately-owned electrical utilities (Tomain, 1997). However, through the 1920s, the drive to achieve lower costs through economies of scale led to greatly increased concentration of the industry. Between 1922 and 1927, over 1,600 privately owned electrical systems disappeared or were taken over (Tomain, 1997), and many of the remaining utilities came under common ownership through holding-company structures. By the end of the 1920s, ten holding companies controlled 75% of the production of electricity in the United States (Bakke, 2016).

Not only was there increasing concentration in the utilities sector, but there was also continuing decline in consumer self-generation. As late as 1922, 25% of the electricity produced in the United States still came from decentralized self-generation (primarily industrial) rather than from electric utilities (Joskow, 1989). However, with continued cost reductions by central station utilities, decentralized self-generation declined steadily over the following decades until accounting for only 3% of electricity production, almost entirely by industrial enterprises, by the 1970s (Joskow, 1989).

This environment in the 1920s and 1930s proved to be the setting for the evolution of the dominant utility business model discussed in this chapter. However, unlike earlier business model transitions, which were driven by changes in the value offering (see the description of Edison's value offering of lighting and electricity in Chapter 4) or the value chain (see the description of the changes to the value chain driven by technological innovation in Chapter 5) the factors driving the transition were most associated with value capture, and established through the regulatory compact.

This new distribution utility business model that emerged in the 1920s and 1930s has largely continued into the early twenty-first century and is represented conceptually in Figure 6.4.

Value Capture

In exchange for the award of a monopoly franchise to the utility, the government, through an appointed regulator, gained the authority to cap the profit that a utility could earn on that franchise. It would do so through Cost of Service Regulation, by setting electricity prices at a level that protected the ratepayer from the pricing power of the monopoly, while enabling the utility to earn a reasonable rate of return on its investment (Tomain, 1997).

Value Offering

As part of the regulatory compact, the utility took on an obligation to deliver a quality, reliable supply of electricity at reasonable cost within that monopoly franchise area (Tomain, 1997). The ratepayer gained assurance of fair treatment from a monopoly supplier through regulatory tests that the utility's investment be "prudent" and "used and useful."

Customer Identification

The regulatory compact provided the utility with monopoly status in an exclusive service territory. As part of that compact, the utility took on "an obligation to serve" all customers within that territory.

Third Dominant Business Model
Centralized Production, Regulated

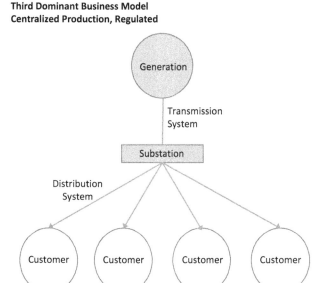

Figure 6.4: Third Dominant Business Model.
Source: Author

Value Chain

The monopoly franchise awarded to utilities eliminated overlapping service territories, enabling focus on standardization, reliability, and efficiency within the utility's service territory. The dominant business model following the 1920s was characterized by fully integrated generation, transmission, and distribution systems with regulated, monopoly franchises. However, as noted later in this chapter, in the latter years of the twentieth century, some jurisdictions revised the regulatory framework governing the generation, transmission and retail segments of the industry.

This description of the utility model that has been dominant for the last eighty years is necessarily high-level, due to the variances that exist between jurisdictions and regulatory environments. Nevertheless, the description certainly is still a fair description of integrated utilities in regulated environments, and for distribution utilities in deregulated environments. The framework provides an excellent structure to describe the strains that the business model is under today, as will be explored in following chapters.

Key Variants of the Dominant Business Model – 1970s to Present

The three decades following the implementation of the Public Utility Holding Company Act in the United States, from 1935 to about 1965, would become known as the "Golden Age" of electricity to historians of the industry (Tomain, 1997). During this period, demand for electricity in the United States doubled every ten years, growing at an annual rate of roughly 7% (Tomain, 1997). Scale economies and new technologies kept real rates flat or declining, and for most utilities, the marginal cost of most new production of electricity throughout the period was consistently less than the average cost of existing production (Tomain, 1997). With the marginal cost of new production typically less than last year's rates, the process of ratemaking was a fairly easy task for regulators, utilities earned healthy and stable returns, and there was little pressure on the dominant business model to change. However, with the energy crisis of the early 1970s, the marginal cost of new production started to exceed the average cost of production for many utilities (Tomain, 1997). The result was a trend of increasing costs to users, reduced profits for utilities, and increased regulatory tension and animosity. As a result, starting in the 1970s, there were significant regulatory changes in some jurisdictions to affect the environment under which utilities operate. However, these changes were primarily focused on generation and transmission areas of the industry, and had only limited, secondary effect on the dominant business model of the electrical distribution business.

There were several waves of regulatory change:

1970s and 1980s: Third-Party Generation

US and Canadian jurisdictions saw the start of the erosion of the fully integrated utility with the introduction of legislation to encourage third-party generation of electricity. In the United States, the enactment of the Public Utility Regulatory Policies Act (PURPA) in 1978 compelled utilities to purchase the output of these third-party generators if their offered price was less than the utility's own avoided costs (Tuttle et al., 2016). In many provinces in Canada, this move away from monopoly control of generation also started through the 1980s, particularly through the development of environmentally beneficial generation sources such as wood waste or small-scale hydroelectric generation (Netherton, 2007).

The 1980s and 1990s: Separation of Transmission and Generation

The 1980s and 1990s saw further erosion of the fully integrated utility in many jurisdictions with the mandated separation of many previously integrated utilities into separate generation, transmission, and distribution companies. This restructuring was driven, in part, by the political change introduced by the Reagan and Thatcher governments in the United States and UK in the late 1980s (McDermott, 2012). The US and UK governments, followed by Australia, Canada, Chile, and others, introduced privatization and deregulation of a range of industries including telecommunications, airlines, natural gas and, of course, electricity (McDermott, 2012). The business of electricity generation was no longer treated as a natural monopoly, and these deregulated jurisdictions introduced various forms of competitive markets to produce electricity, particularly in the production of wholesale power (Tuttle et al., 2016). Regulators enabled the move to market forces by ensuring non-utility generators access to utility-owned transmission systems on a nondiscriminatory basis.

1990s and 2000s: Market Restructuring

In the 1990s, policy makers in some markets moved to further separate the value chains of the former integrated utilities by allowing consumers to chose between electricity retailers, a policy known as "retail choice." In those jurisdictions where it was implemented, non-utility companies were allowed to procure wholesale electricity and sell it to end consumers, often offering alternative retail pricing structures, although customers could typically opt for a traditional regulated rate (Borenstein & Bushnell, 2015). Providing this choice to consumers would typically require utilities to separate their distribution operations from generation and transmission, generally by divesting those functions, or by moving them into separate, arms-length companies. Distribution utilities operating in jurisdictions with "retail choice" were generally responsible for three main activities: operating and maintaining the grid from the distribution substation to the customer meter, customer billing, and acting as a provider of last resort. The provider of last resort in a market operating with retail choice acts as a safety net for the consumer, fulfilling the "obligation to serve" in the event that the retail energy provider is forced to exit the market (Borenstein & Bushnell, 2015).

Deregulation and market restructuring in the 1990s and 2000s did not proceed as smoothly as planned in some jurisdictions. In California, which had been a leader in moving to competitive retail markets, utilities suffered large financial losses and two of the largest utilities in the state went into insolvency. In addition, one of the United States' largest energy suppliers, Enron, went into bankruptcy in 2001 under a cloud of fraudulent transactions and market manipulations. The public concern about the consequences of deregulation slowed or halted the initiative in many jurisdictions (McDermott, 2012). Today, fifteen American states and the District of Columbia[5] operate with

---

5 In the United States, there are fifteen states plus the District of Columbia with varying forms of deregulation of the electricity market, including California, Connecticut, Delaware, Illinois, Massachusetts, Maryland, Maine, Michigan, Montana, New Hampshire, New Jersey, New York, Ohio,

some form of a deregulated market for electricity, while thirty-five states remain under traditional regulation. In Canada, eight of the ten provinces remain regulated jurisdictions, with Alberta and Ontario under forms of deregulation, while a move to deregulation in British Columbia was suspended and ultimately terminated.

Effect of Deregulation on the Business Model

In jurisdictions that have gone through some form of deregulation of their electricity markets, the impacts have been focused on the upstream end of the electric utility (i.e., the generation and transmission portions of the value chain) and, in some jurisdictions, the retailing part of the value chain. However, except for factors such as the addition of digital control capabilities and increasing distributed resources (which will be discussed in the next chapter), the traditional value chain for distribution of electricity from the substation to the customer has remained remarkably constant since the 1930s. It has been said in the industry that if one were to bring Alexander Graham Bell into a modern telecommunications center, he could not begin to understand the workings of the system. However, if you were to show Samuel Insull a modern-day electrical distribution system, he would quickly recognize and understand its core workings and methods of conducting business. The downstream sector of the utility, the distribution of electricity between the transmission substation to the customer's meter, has largely been unaffected by these forces of deregulation. The electrical distribution system continues to be regarded as a natural monopoly and is regulated accordingly (Borenstein & Bushnell, 2015), just as Insull proposed in his 1898 speech.

The next two chapters will describe the factors affecting the utility business model today, with a focus on the electrical distribution utility. Although electricity distribution operates with a business model that has changed little in decades, the observer should not assume that the maturity of the business model implies an absence of forces acting upon it. External factors like resilience management and decarbonization are driving policy makers to place both restrictions and new demands on the utility. Changing technologies in distributed energy are moving parts of the business from a natural monopoly to a highly competitive market. The once-reliable customer relationship is changing to that of both supplier and consumer. Pricing mechanisms are coming under stress, as customers demand unbundled services before the utility can unbundle its rates, and new suppliers move in to provide customers with limited service at lower prices. The pressures on the business model (see Table 6.3) in today's environment are anything but simple.

Pennsylvania, Rhode Island, and Texas, plus the District of Columbia (American Public Power Association, 2020).

Table 6.3: Summary of Third Dominant Utility Business Model – 1920s to 2020s.

Business Model Element	Description
Key Characteristics	Centralized Production. 　Regulated.
Customer Identification	The regulatory compact provides the utility with monopoly status in an exclusive service territory. 　Utility carries an "obligation to serve" all consumers in that monopoly territory.
Value Offering	Provision of an essential service to largely dependent consumers. 　In return for the provision of monopoly franchise, the utility is expected to provide quality, reliable supply of electricity at reasonable cost.
Value Chain	Initially fully integrated. 　Deregulation in some jurisdictions have separated generation, transmission, and retail from distribution operations. However, distribution continues to be treated as a regulated monopoly in virtually all jurisdictions.
Value Capture	Regulated rate of return. 　Volumetric rates established by utility's cost of service. 　Rates represent compensation for a "bundled good," including energy, wires, and other services.

A Final Note Regarding Sam

The stars of the show in the discussion of utility business model innovations have been Thomas Edison, George Westinghouse, Nikola Tesla, and Samuel Insull. Three of these names are well known today. Edison and Westinghouse have remained part of modern folklore for decades, in part due to the enduring legacy of the self-named businesses that they founded. The Tesla name has been rediscovered in popular culture in recent years to achieve rock-star status, due in part to the company named in his honor. However, even though he was a prime architect of two of the three dominant utility business models discussed in this book, Sam Insull's name and the history of the man are little known today.

Insull led a remarkable life. He had a Forrest Gump-like ability to interact with the famous, working alongside a young George Bernard Shaw at Edison's London telephone company, acting as Thomas Edison's right hand for years, and eventually meeting with presidents and prime ministers. In Chicago, he reportedly acted as part of the shadowy "Secret Six" group of concerned businessmen in Chicago who brought Elliott Ness to Chicago to fight organized crime, and funded the investigation that ultimately brought down Al Capone for tax evasion (Hoffman, 2010). As Elliott Ness recalled in his biography, these six men:

> were gambling with their lives, unarmed, to accomplish what three thousand police and three
> hundred prohibition agents had failed miserably to accomplish: The liquidation of a criminal
> combine.   (Ness & Fraley, 1957, as cited in Hoffman, 2010, p. 4)

His was a true rags-to-riches story, starting from the position of a private secretary, he took control of a small Chicago-based utility to became one of the wealthiest and most powerful people in the United States, twice on the cover of *Time Magazine,* controlling railroads and utilities in thirty-eight states, and generating a tenth of the country's electricity. Thomas Edison called him "one of the greatest businessmen in the USA" (Wasik, 2006, p. 3). He kept investing in his companies through the Great Crash because he believed in their viability, lost his fortune, and was tried and acquitted for stock fraud in a case that ran in headlines across America. He was reviled by many, as the 1932 collapse of his holding companies affected the savings of some 600 thousand shareholders and 500 thousand bondholders. He died alone of a heart attack in 1938 in a station on the Paris metro with 84 cents in his pocket, his wallet apparently stolen (Wasik, 2006). Although his death was reported in newspapers across the United States, his funeral in London was attended by only seventeen mourners. He was a remarkable individual with a remarkable story.[6]

---

[6] For those interested in the story of Samuel Insull, see John Wasik's excellent 2006 biography, "The Merchant of Power."

Part 3: **Emergence of a New Business Model**

Part 3 will examine the factors that are driving the development of a new dominant business model for the distribution of electricity.
— Chapter 7 will examine the key factors impacting the current utility business model.
— Chapter 8 will examine current impacts on the "value chain" element of the business model.
— Chapter 9 will examine current impacts on "customer identification."
— Chapter 10 will examine current impacts on "value capture."
— Chapter 11 will examine current impacts on the utility "value offering."
— Chapter 12 will examine the utility business model emerging from these factors.

There are two terms that will be used regularly in this section that should be defined as a reminder for readers less familiar with this subject area (definitions are also included in the glossary).

Distributed Energy Resources: One of the key drivers of change of the utility model discussed in the following chapters, is the growth of distributed energy resources, or "DER." This book follows the definition set out by the New York Independent System Operator, which defines DER as:

> "behind-the-meter" power generation and storage resources, typically located on an end-use customer's premises and operated for the purpose of supplying all or a portion of the customer's electric load, and may also be capable of injecting power into the transmission and/or distribution system, or into a non-utility local network in parallel with the utility grid. These DERs includes such technologies as solar PV (photo-voltaic), CHP (combined heat and power) or co-generation systems, microgrids, wind turbines, micro turbines, back-up generators and energy storage. (DNV GL Energy, 2014, p. 1)

Electric Utility: In previous sections, the business model of the utility examined the firm as a fully integrated entity. However, since deregulation in the 1990s, the business models of generation and transmission have become increasingly separated from the distribution utility business, often existing in separately regulated or deregulated marketplaces. In addition, many of the drivers affecting utility business models that we discuss in the following chapters have their greatest impact on the distribution grid and have only secondary or no impact on generation and transmission. Accordingly, the analysis in Part 3 will focus primarily upon the "distribution utility," and less on the generation and transmission parts of the business. This is consistent with the definition of the US Department of Energy, which defines a utility as a legal entity "aligned with distribution facilities for delivery of electric energy for use primarily by the public" (US Department of Energy, 2018f, p. 249).

https://doi.org/10.1515/9783110714036-009

7 A Business Model Under Strain: Market Disruptions

> Does the grid just become a backup system the way the post office is effectively a backup system for Federal Express and UPS for high-value mail? Or do we actually get to the point where we are tearing down the grid because it's actually not being used at all?
>
> David Crane, Former CEO, NRG (Howland, 2014, p. 1)

Pundits and speakers at conferences like to find a "hook" to describe a phenomenon. The phase used at today's conferences to describe changes impacting the electric utility sector is, "the three D's: decentralization, decarbonization, and digitization." This chapter also describes a fourth factor affecting many utilities searching for financial viability: declining load growth. Although cliched, the terms are a useful framework to describe the big drivers of change moving through the industry. Decentralization describes what is happening, digitization helps explain how, while decarbonization and declining load growth help to understand why change is happening.

Decentralization

The electrical distribution system built in 1882 by Thomas Edison at Pearl Street was based on the idea of centralized power generation, with distribution to consumers through an electrical grid. Later technological breakthroughs enabled larger distribution territories with more diverse customer profiles, which in turn allowed the construction of larger and more efficient generation plants. Although most electricity consumed in the United States was self-generated until 1915 (Granovetter & McGuire, 1998), the cost advantage of grid-supplied electricity was such that self-generation accounted for only 3% of electricity production by the 1970s (Joskow, 1989). Grid-supplied electricity, operating with monopoly protection and large cost advantages, has not had serious competition from self-generation for almost a century. In recent years, however, that has begun to change.

Competing technologies for grid-supplied electricity have been around a long time. Solar cells have been used since the 1950s to power satellites and space stations, and solar panels were famously installed atop the Carter White House in the 1970s at the height of the energy crisis (Himmelman, 2012). But compared to grid-supplied electricity, solar generated electricity has historically been expensive, keeping implementation rates low. However, prices of these competing technologies have been declining sharply. Between 2010 and 2018, the installed cost per kilowatt of a residential solar electricity system fell by over 60% in the United States (Fu, Feldman, & Margolis, 2018) (see Figure 7.1). This is driven by a reduction in all elements of the cost of an installed system, but particularly by reductions in the cost

https://doi.org/10.1515/9783110714036-010

of crystalline-silicon photovoltaic modules (note that these modules form only part of the cost of an installed system), which declined in price from $79/W to $0.37/W between 1976 and 2016 (Bloomberg New Energy Finance, 2018). This cost decline reflects a remarkable 28.5% learning rate (i.e., the percentage cost reduction for each doubling of cumulative global capacity) over that forty-year period (Bloomberg New Energy Finance, 2018).

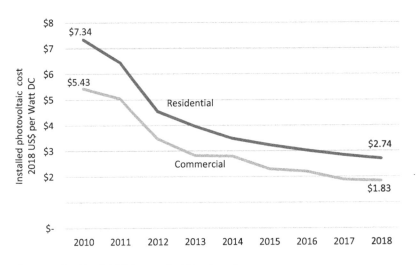

Figure 7.1: Residential PV System Cost Benchmark.
Note: Data for "Residential PV System Cost Benchmark" sourced from the NREL US Solar Voltaic System Cost Benchmark: Q1 2018 (Fu, Feldman, & Margolis, 2018)

As a result of these reductions, the installed cost of an average-sized residential photovoltaic (PV) system in the United States dropped by over half in ten years (SEIA, 2020), to about $19,000 in 2020 (Feldman, Ramasamy, Fu, Ramdas, Desai, & Margolis, 2021). Driven by these cost decreases and government incentives, the installation of residential solar photovoltaic systems has seen remarkable growth in the past decade (see Figure 7.2). Between 2009 and 2019, the US residential photovoltaic market grew at a compound annual growth rate of 44% to 2.3 million homes (US Energy Information Administration, 2020a). Solar PV (including utility scale PV) still only accounted for about 2% of total US generation in 2018 but is forecast to grow more than 7-fold to 14% of total generation by 2050. Of this, almost half (47%) the total solar PV electricity generated in the United States is projected to be generated by small-scale, decentralized systems such as rooftop solar (US Energy Information Administration, 2018). These decentralized sources of electricity, located in proximity to consumption, are part of a growing portfolio of distributed energy resources (DER) that will represent an increasingly legitimate option to grid-supplied electricity in many jurisdictions.

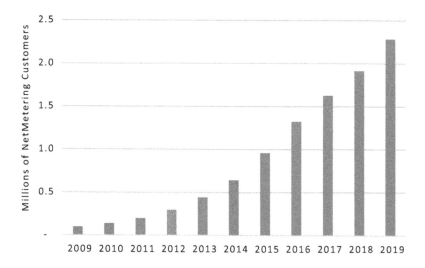

Figure 7.2: Residential Solar PV Penetration ER.
Note: Chart data sourced from the Annual Energy Outlook 2020 (US Energy Information Administration, 2020a)

European deployment of solar generation varies widely between member countries. Germany, where decentralized solar has seen considerable investment due to a generous feed in tariff,[1] has the largest deployment in Europe, with 590 watts of installed solar capacity per capita (EurObserv'ER, 2020). Neighboring Poland, on the other hand, with similar solar potential but different government priorities, has only 35 watts of installed capacity per capita (EurObserv'ER, 2020). Nevertheless, subcommittees of the European parliament are developing plans to support further rollouts of solar. A 2016 study found that European rooftops have the potential to produce almost a quarter of the EU's 2016 consumption of electricity, with two-thirds of that a lower cost than electricity prices then in effect (Bodis, Kougias, Jager-Waldau, Taylor, & Szabo, 2019).

Canada has seen a slower rate of adoption than the United States or Europe, with solar representing less than 0.5% of all generation across the country (National Energy Board, 2018). One reason for this lower rate of adoption is the absence of government supported incentives found in many parts of the United States and Europe. An indicator of the significance of these incentives is that the only Canadian province ever to fund a significant solar incentive program, Ontario, has 98% of Canada's PV capacity (National Energy Board, 2018). Canada's northern geography is sometimes noted as a second reason to explain a slower DER rate of adoption, relative to parts of the United States or Europe. Although it is true that Canadian conditions for PV may not be as favorable as those in California or Arizona, it is also

---

1 See the glossary for a definition of "feed in tariff."

true that most Canadian cities are found at more southerly latitudes than many European jurisdictions with much greater solar penetration, such as Germany or Denmark. For example, the Canadian city of Montreal lies at a latitude similar to Milan, Calgary to Dresden, and Toronto to the Spanish city of San Sebastian. Decentralized solar will grow significantly in Canada if prices continue to fall and if governments use financial incentives to support decarbonization imperatives (National Energy Board, 2018).

A further factor that has the potential to add to the growth of distributed resources is the significant decline in the cost of battery storage. In 2014, researchers using an average of industry and government sources, predicted a 50% decline in battery costs over ten years, from approximately $700/kWh in 2013 to $350/kWh by 2023 (Bronski et al., 2014). In fact, the decline in battery costs have blown through those forecasts, falling to an average cost of $137/kWh by 2020, an 89% decline over the previous ten years (BloombergNEF, 2020a) (see Figure 7.3). Battery prices are predicted to continue to decline, reaching $101/kWh by 2023 based on continuing advances in materials chemistry and scale economies, and $58/kwh by 2030 based on continuing technological advances in areas such as solid-state batteries (BloombergNEF, 2020a).

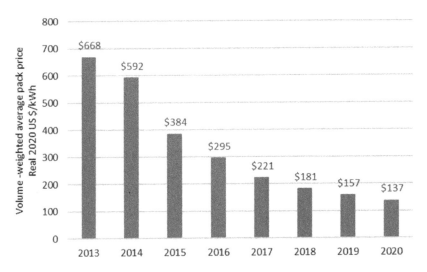

Figure 7.3: Volume-Weighted Battery Pack Price.
Note: Chart data sourced from BloombergNEF (BloombergNEF, 2020a)

It is important to note that storage is not, itself, a source of generation, and its current installations on the grid have been largely used to manage frequency regulation (Spector, 2020a). However, the growing value of storage is that it enables the consumption of electricity generated by a renewable energy source, like a solar panel or wind turbine, to be deferred to a later time. When combined with an inherently variable resource, storage can add significant value to the grid by allowing better utilization of assets in the system (Spector, 2020a), and does so in three main ways. First, it makes variable renewable generation resources more productive, by allowing electricity that would otherwise be curtailed and lost (e.g., energy from a solar system at mid-day when demand is low), to be stored and used later (Mallapragada, Sepulveda, & Jenkins, 2020 as cited in Spector, 2020a). Second, storage can be used instead of fossil-fuel based generation plants that would otherwise be used in periods of peak demand, thereby improving the carbon footprint of the electrical system (Mallapragada, Sepulveda, & Jenkins, 2020 as cited in Spector, 2020). Third, storage can be used to manage transmission capacity, by holding back power from transmission in periods of system congestion, and releasing it when the transmission system has greater capacity (Mallapragada, Sepulveda, & Jenkins, 2020 as cited in Spector, 2020a). This capability will allow the avoidance or deferral of investments in incremental transmission capacity.

Given their remarkable cost declines, the installation of distributed generation and storage is forecast to grow significantly over the next decade (US EIA, 2020a). How will this affect the utility business model? When these generation and storage assets are located on the utility's side of the customer meter, as grid-attached, utility-owned assets, the effect on the utility business model is limited. If the utility is able to include these assets in the rate base[2] and be compensated for investing in them, the utility's value offering and value capture remain largely unchanged. In fact, the customer may very likely be unaware of the new generation or storage assets.

However, if the distributed energy resources are held on the customer's side of the meter, as is quite common in many jurisdictions, there could be a significant impact on the utility business model, particularly as the penetration of these resources increase. Many aspects of the business model, including the definition of the value chain, the services offered to the customer, the method of compensation for those services, and even the definition of the customer, are altered. These strains on the utility business model will be discussed in the following chapters.

Decarbonization

The generation of electricity is a significant source of GHGs, a legacy of its long-standing reliance on fossil fuels. In Europe, which has been a leader in the implementation of renewable energy, fossil fuels still account for 40% of electricity generation (European Commission, 2020). Even in Canada, a country blessed with a large hydroelectric base in many provinces, fossil fuels account for 18% of the country's

---

2 See the glossary for definition of "rate base."

generation (Government of Canada, 2018), making the electricity industry the country's fourth largest source of GHGs (Government of Canada, 2018). In the United States, although there has been a marked decline in coal-fired generation in the past decade, most of that decline has been offset by an increase in generation fueled by natural gas. Despite highly visible investments in renewables, in the thirty years from 1990 to 2019, the share of fossil fuel-based generation in the United States declined only from 69% to 63% (see Figure 7.4) (US EIA, 2021c). The generation of electricity is still the second largest producer of GHGs in the country, accounting for a quarter of total US GHG emissions (US EPA, 2021).

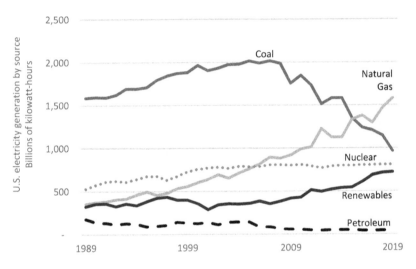

Figure 7.4: US Electricity Generation by Major Energy Source, 1989–2019.
Note: Chart data sourced from US Energy Information Administration (2020b).
Monthly Energy Review, Table 7.2a, March 2020

Although the electrical grid occupies a central role in the production of GHGs, it is also an important tool to support policy makers pursuing the reduction of GHGs. It does so in two main ways. First, by enabling a change in energy sources, from fossil-based technologies to those with minimal GHG impact such as decentralized wind or solar, and second, by enabling the electrification of other sectors, such as transportation, a leading source of GHGs in most economies.

Commitments to reduce GHGs by governments around the world are driving changes in the electrical grid. The United States has accelerated its pace in this area, as a new US Administration has committed to a carbon free electrical grid by 2035. If implemented, the initiative will mark a sharp change in direction, as the most aggressive prior plan at a state level, in New York, previously had a target date in 2040 (Baker & Kaufman, 2020). The initiative is ambitious, since to achieve even a level of 75% noncarbon sources will require doubling current levels of nuclear and renewables (US EIA, 2020b). The United States' major trading partners in Europe, Canada, and Japan have similarly set targets for reductions in carbon in the generation of electricity, although often with less challenging deadlines. In Canada, for example, the imposition of a federal carbon tax and provincial initiatives to retire coal-fired generation plants are part of a plan to achieve net-zero emissions by 2050 (Government of Canada, 2020). Europe similarly plans to reach carbon neutrality by 2050 with substantially all electricity being generated from carbon free sources by that date (European Commission, 2020). Japan and the UK have joined the European Union in the net-zero target by 2050, with China targeting peak carbon emissions in 2030 and carbon neutrality by 2060 (Tsukimori, 2020).

Internationally, corporations are also making significant commitments to reduce their carbon footprint. More than 260 major corporations, including Google, IKEA, Apple, Mars, and Citibank, have already made long-term commitments to achieve 100% renewable energy consumption, with 75% of those targeting to do so by 2030 (Re100, 2020). IKEA, for example, has installed rooftop solar at over 90% of their American stores, while Walmart has installed 145 MW of distributed rooftop solar at 500 of its facilities across twenty-two states (Roselund, 2018). Every week, over six million Americans shop at a Walmart store running on rooftop solar (Fehrenbacher, 2016).

How does this growing investment in renewables impact the utility business model? First, utilities will be faced with increasing generation of renewable resources, including distributed renewable resources. To the extent that renewable resources are utility scale, they will largely fit withing the existing business model. However, to the extent that the resources are distributed, and potentially owned by the customer, they could have a significant impact on the utility business model. Second, the effort to decarbonize the economy, and the transportation sector in particular, will require utilities to support infrastructure that can charge fleets of new electric vehicles. As these vehicle volumes scale up, this will represent a significant new product offering for utilities and could have significant impact on the value chain at the level of the local distribution system. As discussed in following chapters, both distributed renewables and electric vehicles have the potential to significantly impact the utility business model, requiring the deployment of new grid assets, tariff schemes, channels of communication to the customer, and service offerings.

Digitization

For the last decade, distribution utilities across North America have spent billions implementing smart grid[3] technologies to help manage their electrical systems. The US Department of Energy estimates that $32.5 billion has been spent in the United

---

3 See the glossary for definition of "smart grid."

States between 2008 and 2017 specifically on smart grid programs (Campbell, 2018). A useful indicator of the progress of these digitization capabilities is the deployment of smart meters[4] (see Figure 7.5).

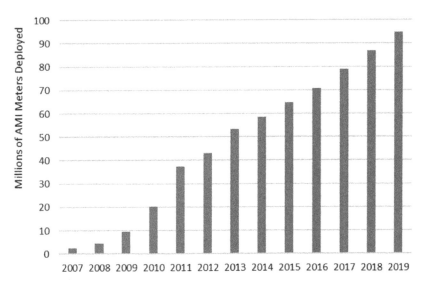

Figure 7.5: Advanced Metering Deployment – United States.
Note: Data for Advanced Metering Deployment from: "Advanced Metering Count by Technology Type," US Energy Information Administration (US EIA, 2021b)

Although the smart meter is only a small, inexpensive device at the traditional terminus of the regulated electrical distribution system, its existence is an indicator of the significant infrastructure that has been built to support it. If a smart meter is operating at a customer's house, it typically means that the utility has invested hundreds of millions of dollars in distribution equipment, communication infrastructure, and IT systems required to manage communication with that meter, and with the customer.

By the end of 2020, utilities in the United States had deployed over 94 million smart meters representing about 61% of the 154 million meters in the United States (US Energy Information Administration, 2021b). By comparison, the European Union has converted about 72% of its meter population to smart meters by the end of 2020 (European Commission, 2021), while the UK has converted about 40% of the meter population with plans to convert substantially all of the population by 2024 (Holder, 2020).

---

4 See the glossary for definition of "smart meter.".

These smart grid investments in controls, IT systems, telecommunications systems and automation have greatly expanded the capability of distribution utilities to manage their grid with greater reliability and efficiency. Some of this investment will not fundamentally alter the business model but may allow it to operate more efficiently or effectively. For example, smart grid investments in feeder and substation automation and improved outage management will allow the grid to operate more efficiently and effectively, but do not fundamentally alter the business model of the traditional distribution utility. However, some smart grid investments have also laid the technical foundations to alter the utility business model. For example, smart grid investment can enable utilities to manage distributed energy resources on the customer side of the meter, extending the traditional utility value chain. It can open new communications channels to the customer and provide new value to the customer with access to timely data on electricity consumption to support more efficient electricity usage. Smart grid investments will enable the utility value chain to better respond to the growing demand for the charging of electric vehicles. These elements, and others discussed in the following chapters, can enable substantial changes to the business model.

Declining Load Growth

Electricity is more important than ever in our day-to-day lives. Our homes are packed with electrical devices, and our always-connected lifestyles require freshly charged batteries of the devices in our pockets. However, despite this proliferation of electrical devices, the growth in the consumption of electricity in most industrialized nations has declined considerably in the last few decades. For example, after growing at an annual rate of over 7% through the 1950s and 1960s, growth in consumption in the United States declined to an annual rate of about half a percent per year through the first decade of this millennium (US EIA, 2021b), and about two tenths of a percent through the 2010s (US EIA, 2021b) (see Figure 7.6).

The rate of growth in electricity production is even slower when viewed on a per capita basis in most developed economies. According to the most recent data from the World Bank, in the period from the end the recession of 2008 to 2014, the consumption of electricity per capita actually fell by close to 5% in each of the United States and Canada, by 6% in the European Union, and by 10% in Japan (World Bank, 2021a). One of the main drivers of the decline is the continuing shift in developed economies from manufacturing and resource extraction to a less energy intensive service-based economy (Hirsh 2011). Today's developed economies are much less energy intensive than in the past. In this millennium alone, from 2000 to 2014 (the most recent year for which data is available), the amount of energy required to produce a constant dollar of GDP fell by 38% in each of the United States and Canada, by 28% in the European Union, and by 33% in Japan (World Bank, 2021b).

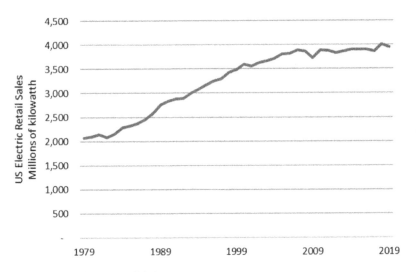

Figure 7.6: US Electric Retail Sales.
Source: US Energy Information Administration (US EIA, 2021a)

The decline in per capita consumption of electricity also arises from the increased efficiency of items like factory equipment, lighting, home appliances and buildings, often driven by government standards and building codes (Hirsh, 2011). A simple efficiency measure like changing to a more efficient lighting system may seems like a small act on its own, but the aggregate impact across a whole electrical system can be significant. The contribution of this type of energy efficiency to load decline was recognized by this utility executive:

> well, let's think about the math for a minute. So, generally speaking if you look at [our] utility sales, about thirty percent of it is for lighting.. . . When you re-lamp a commercial building with LEDs, you are saving eighty percent of the energy that you were consuming before.. . . So that's twenty-four percent of all the sales that you had with that customer, now are gone. If you take that twenty-four percent, . . . and you play that across the country, that's about what everybody will be off, and that's about how much we will be off on sales.
>
> Utility executive, author's research session C3.[5]

Although the slowing of growth in consumption is noticeable across all major sectors of the economy, the impact has been strongest in the industrial sector. The industrial sector was the largest consumer of electricity for decades, until the mid-1990s when

---

5 As part of an academic research project conducted by the author, interviews sessions were held with executives at a range of utilities across North America, contributing to a published doctoral dissertation in 2019. To ensure a frank discussion of issues affecting these companies, confidentiality of each interviewee's name and employer was maintained, except to academic reviewers. This and following chapter will utilize quotations from these interviews to illustrate changes that are occurring in the utility business model.

growing efficiency and the shift to a services economy caused growth to flatten. In the twenty-five years between 1994 and 2019, American gross domestic product grew by 88% in real terms (US Bureau of Economic Analysis, 2021), while the consumption of electricity by the industrial sector was virtually unchanged (see Figure 7.7).

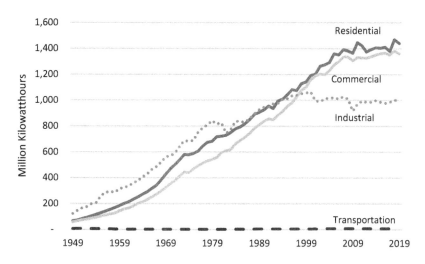

Figure 7.7: US Electricity Sales by Sector (1949–2019).
Source: Data for US Electricity Sales by Sector from "Electricity End Use" US Energy Information Administration (US Energy Information Administration, 2021a)

Forecasts for the next few decades show modest growth in consumption of electricity, although nowhere near historic levels. US consumption is forecast to grow at an annual rate of about 1% between 2020 and 2050 (US Energy Information Administration, 2018), while Canadian growth rates are forecast similarly at 1% between 2017 and 2040 (Natural Resources Canada, 2018), both rates much lower than the forecast growth of their respective economies.

Some of this declining load will be offset by increased electrification of the transportation system. As an exercise to understand the upper bounds of the potential impact of growth of electric vehicles, one nonpeer reviewed study estimated that if all 251 million light vehicles in the United Sates were replaced with electric vehicles, it would add 25% to the forecast demand for electricity (Walton, 2021). However, most current forecasts reflect a much more modest impact from the electrification of the transportation sector. For example, the 2020 reference forecast of the US Energy Information Administration indicates that by 2050, the transportation sector would still account for less than 2% of total electrical consumption. The forecast notes that "both vehicle sales and utilization (miles driven) would need to increase substantially for EVs to raise electric power demand growth rates by more than a fraction of a percentage per year" (US EIA 2020a, p. 64). Other forecasts show somewhat greater impact of electric vehicles with forecasts between 2% to 3% (Bloomberg-NEF, 2020b) of total demand by 2030. However, in the United States and Europe much of this growth from electric vehicles is forecast to be largely offset by continuing reduction in consumption from energy efficiency (BloombergNEF 2020b). In all scenarios, forecast growth in electric vehicles is subject to significant political uncertainty.

As tempting as transportation may appear as a source of load growth, research has shown that forecasts of growth in electricity demand have often tended to be higher than growth rates actually achieved. A 2016 study that retrospectively studied the load forecasts of twelve Western US utilities between the mid-2000s and 2014, found that all but one of these utilities consistently overestimated the growth of electricity consumption and peak demand (Carvallo, Larsen, Sanstad, & Goldman, 2018).

Although current forecasts vary greatly in assessing the impact of electric vehicle charging at the aggregate level, there is much greater agreement that the impact could be much greater at the level of local distribution systems. The load profiles of a local system can change quickly as fast charging stations are installed and as neighborhood pockets adopt electric vehicles. This will be discussed further in Chapter 8.

Implications of Declining Load Growth

Flat or declining sales can be tough for any company. For a high fixed cost regulated utility, the effects can be especially difficult. In an environment of ever-expanding load, an electric utility can spread its fixed costs over a wider base to reduce unit costs to the customer, allowing it to reduce rates while continuing to invest in an expanding electrical system, and earning a regulated but reasonable rate of return on its investment (Tomain, 1997). However, when loads start to decline, fixed costs are spread over a narrower base, and average unit costs increase, resulting in rate increases. As higher prices this year result in reduced demand next year, the cycle continues and repeats in following successive years. As unit costs increase, rates increase, and load volumes again decline. In the industry press, this process widely has been widely referred to as "The Utility Death Spiral" (Graffy & Kihm, 2014, p. 2). Under this scenario, utilities are unable to reduce their largely fixed cost structure as quickly as load decreases, resulting in those fixed costs being carried by fewer and fewer customers with higher and higher rates.

> It (declining load) is a big concern. For us also. It really is . . . And with declining load growth, that's declining revenue, typically. And we still have, just like everybody else, aging infrastructure. We also have the need to make new, incremental investments, and so forth. And the financial impact of that is, the only way to really fund that, is to raise rates.
>
> Utility executive, author's research session C1

Competition from distributed resources in future decades is identified as one of the causes of the lower growth forecast for grid-supplied electricity. The US Energy Information Administration, in its most recent forecast, noted that the growth forecast for grid-supplied energy "would be higher if not for significant growth in generation from rooftop photovoltaic (PV) systems, primarily on residential and commercial buildings, and combined-heat-and-power systems in industrial and some commercial applications" (US EIA 2020a, p. 64). Although declining growth of grid-supplied electricity does not have a direct impact on a firm's business model, it will have an indirect impact as declining profitability will motivate firms to seek new business models. This further disruption in the traditional utility business model is the subject of following chapters.

8 The New Business Model: Decarbonizing the Value Chain

But it is not designed, as are none of the other electric utility distribution systems, to handle large volumes of these [distributed] resources. This is not a system that can handle two-way power flows very well. So, the key issue at the first layer is, 'how to redesign and re-architect the system to handle a much more distributed set of resources operating on the system.'

Utility executive, author's research session E4

There have been some limited changes to the traditional utility business model over the past several decades, particularly in jurisdictions that deregulated electricity generation in the 1990s and early 2000s. Nevertheless, the core value chain for delivery of service to the customer, of generation-to-transmission-to distribution-to-customer, has been quite stable for most of the past century. However, in the past decade, several factors have emerged that may significantly alter the way in which this value chain operates, including the introduction of digital control technologies, the declining cost of decentralized energy resources, and efforts to decarbonize our economy. This chapter will look at some of those key factors that are impacting the utility value chain.

As distributed energy resources (DER) achieve greater market penetration, managing this new source of supply will be challenging within the infrastructure of the traditional distribution utility. An example of the challenges facing distribution systems with high levels of DER was provided in the summer of 2018, as the city of Los Angeles experienced rolling power outages affecting tens of thousands of customers during a particularly intense heat wave. Some politicians delivered a knee-jerk attribution of blame to inadequate generation, calling for investment in new traditional, centralized, generation capacity (Shellenberger, 2018). However, a closer reading of the situation revealed that the likely source of the problem was not inadequate central generation plants, nor a constrained transmission system. In a city that has one of the highest rates of penetration of DER in North America, the operational failures were found to originate with overloaded distribution circuits and equipment (Bade, 2018). The issue was not traditional centralized generation or transmission. Instead, it was the stress placed on the distribution system in a period of high demand and high output to the grid from distributed resources, such as rooftop solar. The distribution system had reached a point of operational instability (Bade, 2018).

Integration of increasing amounts of distributed resource into a traditional utility value chain will be challenging for two main reasons that have to do with the very nature of these resources: their intermittency, and their dispersion (Ramchurn, Vytelingum, Rogers, & Jennings, 2012). Adding to the complexity and urgency of this integration over the next decade, the utility value chain is forecast to be at the center of enabling the shift of large portions of the economy, particularly transportation, from fossil fuels to electricity generated by renewable resources. Nevertheless, despite

https://doi.org/10.1515/9783110714036-011

the challenges, there are also opportunities to capture value from the integration of these new resources with the existing grid, and these will be discussed in the following sections.

Intermittency of Supply

But the thing that really shook our engineers [about distributed energy resources] was the ramp-up speed and the ramp-down speed. And it was within . . . minutes that they would go from flat-out to zero. And from zero to flat-out. . . . But this type of fluctuation that you get from a [distributed energy resource] is extremely difficult to manage.

Utility executive, author's research session A3

Power system engineers learn early in their education that the most important aspect of managing an integrated electrical system is that supply and demand must be kept constantly in balance. In a traditional one-way distribution system, this typically requires managing the supply of generation to meet fluctuations in demand. Even in a traditional electrical system with centralized power plants and one-way distribution of electricity, this can be challenging. For example, after losing to Germany in a shoot-out in the 1990 World Cup, unhappy soccer fans across England turned for solace in a familiar comfort, a nice cup of tea. An estimated 1.1 million tea kettles were turned on across the country within seconds of the English ball going wide in the shoot-out, resulting in an 11% surge of demand on the country's National Grid (BBC News, 1998). Fortunately, operators on the English National Grid were prepared with power supplies in reserve, as are other operators around the world in similar situations, ready for demand fluctuations to deliver additional electricity into the grid when needed.

Supply sources might come out of service for maintenance, or transmission lines might be hit with an unplanned outage, but generally the traditional supply chain operates within a highly predictable range of dispatchable[1] output. However, distributed energy resources based on solar and wind are intermittent by nature, and the task of managing an electrical system that includes high levels of these variable resources involves not only managing fluctuations in demand, but also requires much more attention to fluctuations in supply. The electricity that these resources supply can rise or fall rapidly as wind gusts or the sun disappears behind a cloud. This task is complex enough in a system with one-way flow of electricity from centralized generation to the customer, but the intermittent nature of DER makes matching of supply and demand much more difficult, with flows of electricity that continuously change in voltage and direction (Ramchurn et al., 2012).

---

1 See the glossary for a definition of "dispatchable electricity."

The "Duck"

Even within the time frame of a single day, the intermittency of electricity generated by distributed resources as the sun rises and falls can be highly predictable, but can still make management of the grid tremendously challenging. A well documented example has been the changing shape of the load curve in the state of California over the past decade. As early as 2008, forecasters (Denholm, Margolis, & Milford, 2008) were predicting that the growth of distributed energy resources, and rooftop solar in particular, would dramatically change the daily load curve of the state's electrical system. In 2013, the California Independent System Operator (ISO) translated this forecast impact into the famous "duck curve" (see Figure 8.1), so named due to the resemblance of the profile to a duck (California ISO, 2013). The duck curve maps the forecast change in the net load on the grid on a typical spring day. The net load is calculated by starting with the forecast load, but subtracting the forecast production from variable distributed generation resources, such as rooftop solar or wind (California ISO, 2016).

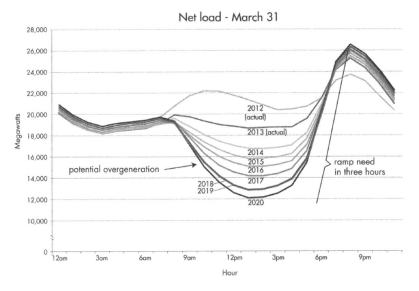

Figure 8.1: Forecast California Duck Curve, typical spring day, published 2013.
Source: California ISO, 2013. Licensed with permission from the California ISO. Any statements, conclusions, summaries or other commentaries expressed herein do not reflect the opinions or endorsement of the California ISO.

The net load curve starts with consumption rising modestly in the morning (outlining the duck's tail), before falling as distributed solar electricity starts to come online, displacing other sources of generation (outlining the duck's belly). Then. In the late afternoon and early evening, as consumers come home from work and solar generation starts to diminish, net load ramps up quickly before diminishing in the late evening (outlining the duck's neck and head). The growing year-after-year change in the curves reflects the increasing penetration of solar resources and their growing impact on the grid on a typical spring day. In fact, the change in shape of the load curve predicted in 2013 has been fairly accurate, although the changes in load shape have arrived about four years earlier than this 2013 forecast, driven by greater than forecast growth of rooftop solar installation (CAISO, 2018).

The Challenge

The changing profile of the curve poses two key challenges to grid operations. The first is the potential for overgeneration during daytime hours, when generation from distributed solar is high and load is relatively low. This could lead to expensive curtailment of baseline gas, coal, or nuclear plants, which is terribly inefficient for plants that have been built to run at levels of relatively constant production, and not to rapidly ramp down or start up. Alternatively, it could force the utility to curtail output of solar generation, which limits environmental benefits and, with wasted electrical output, fails to capture the potential value of the distributed assets (NREL, 2018).

The second challenge of the increasingly exaggerated shape of the duck curve is the evening ramp-up, when generation from distributed solar diminishes while customers start evening activities in their homes, and load on the system is increasing. This requires generation resources to be available to come online rapidly, a difficult and technically challenging task, made more difficult by the increasing slope of the ramp-up over the past several years. By 2016 the California system operator managed its steepest three-hour evening ramp of 10,091 megawatts (MW), an increase of 62% from 2011, and by March 2018 the ramp-up was 14,777 MW (CAISO, 2018) and is forecast to grow to over 18,600 MW by 2021 (NERC, 2020). To give context to the size of this three-hour ramp in the California system, keep in mind that the nameplate capacity of British Columbia Hydro's hydroelectric system, one of the largest in North America, is about 11,921 megawatts (BC Hydro, 2019).

Addressing the Challenge

To be able to manage these changes in the load curve, system planners have taken two broad approaches: "fattening the duck" and "flattening the duck."

"Fattening the duck" requires investments to increase the flexibility of the grid, to enable the grid to respond to continued growth in the belly of the duck (i.e., allowing distributed resources to displace more baseload) without risking the stability of the system (Denholm, O'Connell, Brinkman, & Jorgenson, 2015). This includes investments to make baseload generation plants more flexible, to allow them to increase or decrease load more rapidly (Cochran, Lew, & Kumar, 2013). It also includes regional interchange, by increasing the ability to transmit or draw electricity from other jurisdictions, through increased transmission capacities. This interchange between regions can help to reduce the impact of renewables intermittency by connecting larger energy markets having different supply and load profiles, for example, linking regions with evening wind resources with markets having a midday solar generation surplus (Denholm et al., 2015).

"Flattening the duck" acts to shrink the belly of the duck, by shifting supply and demand patterns to a different part of the day. This could include programs such as time-of-use electricity pricing to alter demand, making electricity more expensive in critical periods, or demand response[2] programs to incent customers to shift consumption to off-peak periods. It could also include investments to shift supply. For example, some jurisdictions are providing financial incentives for some solar panels to face west rather than south, in an effort shift solar production to later in the day when demand increases (Kosowatz, 2018). Alternatively, jurisdictions may consider energy storage technologies such as batteries or pumped hydro to store surplus mid-day supply for later consumption (Denholm et al., 2015).

Historically, battery storage has seen only limited usage to shift supply due to cost disadvantages relative to electricity off the grid 2024 (NERC, 2020). However, with the declining cost of battery technology, and the increasing value to engage in electricity arbitrage in some jurisdictions (i.e., storing energy when it is cheap midday, and selling it later in the day when it is in in greater demand and prices are higher) battery installations are expected to grow substantially in the next decade. The NERC 2020 reference case[3] (see Figure 8.2) forecasts battery storage and hybrid storage (which combines storage with a source of generation, such as solar or wind) to grow from low levels in 2020, to over 45,000 MW by 2024 (NERC, 2020). The role of electric vehicles in this growth could be significant, not only as a potential source of battery storage to feed on to the grid, but also as a load that can be managed in terms of magnitude and timing.

Although it has been most extensively documented in California, the duck curve is migrating, and challenging grid operators around the world. Similar versions of the duck curve are analyzed by planners wherever distributed renewable generation has made substantial inroads, from New England (Spector, 2018) to South Korea (Kim, Kim, Kim, & Cho, 2020) to Australia (Kosowatz, 2018). In Hawaii, for example, where rooftop solar accounts for nearly one tenth of the generation capacity of the state

---

2 See the glossary for definition of "demand response."

3 NERC is a regulatory authority whose mission is to assure the reliability of the bulk power systems in North America. The data used for NERC's 2020 Long-Term Reliability Assessment "captures virtually all electricity supplied in the United States, Canada, and portion of Baja California Norte, Mexico" (NERC, 2020, p. 4).

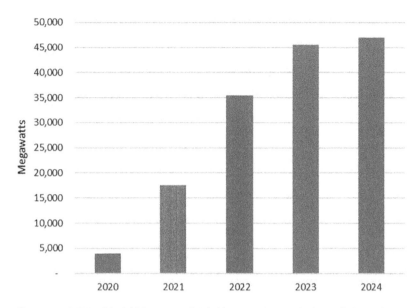

Figure 8.2: NERC-Wide Grid Battery and Hybrid Generation – Existing and Planned.
Source: Data from NERC 2020 Long-Term Reliability Assessment (NERC, 2020). Includes battery storage and hybrid storage (which combines storage with a source of generation).

electric utility, planners have adopted their own version of the curve, nicknamed the "Nessie Curve" after the legendary Scottish Loch Ness Monster (St. John, 2016).

Increased intermittency has increased the complexity of managing the value chain. To manage this complex environment, it may become necessary for the utility to develop control capabilities beyond its traditional boundary of the customer meter. To be able to continuously match supply and demand in a high DER environment, the utility needs some form of limited control of the electricity supplied by the customer's DER (e.g., by controlling a smart inverter[4] on a rooftop solar installation), or of the customer's demand (e.g., by managing curtailable load on air conditioning). The value of this control capability will only increase with growing adoption of electric vehicles, first by providing a potential form of distributed storage (from the vehicle's battery) that can act as a source of supply to the grid, and second, by providing a significant source of curtailable load (Ramchurn et al., 2012).

Dispersion of Resource

So, if you take analogies to water, and trees, if you think of each substation as a tree, then where electricity leaves the station is a branch, and where it leaves the station, the branches

---

4 See the glossary for a short definition of "smart inverter."

are the thickest, because that's where the power has to leave. And it keeps branching off, and each time it branches off, it gets a little thinner, and by the time it reaches the furthest out it is getting very thin where it is weakest. And "weakest" is a bad word, but by design, you don't need big heavy conductor out there because you are moving power to just a few customers versus thousands or hundreds. Now, if distributed generation starts to pop up at the end of those branches, you have a real problem, because it's not designed to move power from that direction.   Utility executive, author's research session E1

In addition to being intermittent, a second challenge posed by distributed resources in the management of the grid is that they are, by definition, distributed. Instead of managing the one-way distribution of electricity sourced from large, centralized power plants, utilities in environments with high penetration of distributed resources must now do so in concert with thousands, or hundreds of thousands of small power generators distributed across the grid. For example, at the end of 2019 the state of California achieved a target it had set years earlier of over one million rooftop solar installations (Roth, 2019). These are net metered[5] accounts, and each one is a potential source of electricity to the grid (see Figure 8.3) to be managed in concert with the state's central coal, nuclear and gas generation plants.

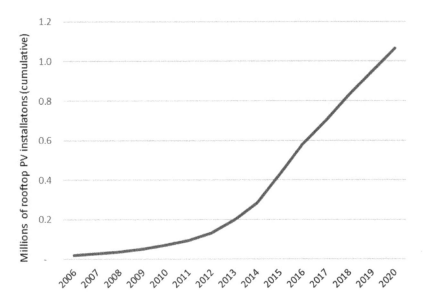

Figure 8.3: Growth of California Rooftop Solar PV Installations.
Source: Data from California Energy Commission (2020). Represents solar PV Installations connected by net energy metering to California's investor owned utilities, PG&E, SCE, and SDG&E.

---

5 Net metering will be discussed in Chapter 10. See the glossary for a short definition of "net meter."

The Challenge

The traditional utility architecture has been refined over decades to efficiently manage the one-way flow of electricity from a handful of generation plants to the consumer. Traditional one-way power systems can continue to operate effectively when the adoption of DERs is quite low. However, these one-way systems simply have not been designed to accept significant volumes of two-way flow. Imagine moving from a traditional environment of centralized production, to one that still includes central generation plants, but now also includes over a million potential sources of supply at the end of the supply chain, adding a multi-directional flow that continuously changes in direction and magnitude (Ramchurn et al., 2012).

The operation of such a supply chain requires new operational procedures and controls systems, and the increasing penetration of DERs will threaten the ability of a traditional distribution network to maintain stable voltages, to ensure worker safety, and to control the risk of cascading power outages (Ramchurn et al., 2012). The two-way flow of electricity will require investment in new hardware to replace fuses, transformers, conductors, and other devices that were never designed to accommodate such an environment. The complexity of the system, and the continuing requirement to maintain stable voltages and frequency, will require new control systems that are able to act much more autonomously. These systems will need to support human supervision but not require direct human control, as the grid will incorporate certain abilities to act autonomously, to self-diagnose problems, and to institute remedial actions. (Ramchurn et al., 2012).

The Opportunity

The task of developing the capability to manage significant penetration of distributed resources will be costly for utilities. However, investment in the integration of distributed resources can also provide utilities with opportunities to defer, reduce or eliminate the upgrade of conventional transmission and distribution assets. Distributed resources are inherently located close to the point of energy consumption, and therefore can be used to alleviate capacity or resilience issues of a distribution network serving a particular group of customers. This type of investment in distributed resources has become known as a non-wires alternative (NWA) (not to be confused with the influential American hip hop group of a similar name). One example of a successful NWA was demonstrated by Duke Energy's Mount Sterling project in 2016, which invested in a solar microgrid with storage as an alternative to upgrading a remote 4-mile (about 6.5 km) distribution line in North Carolina's Great Smoky Mountains National Park (Wood, 2016). In addition to cost saving, the project also allowed the utility to decommission the line and return the right-of-way to park wilderness. Another example, and one of the largest and best-known NWA project, is New York's "Brooklyn-Queens

Demand Management" project, which used a $200 million dollar investment in customer-sited generation, storage, and energy efficiency to defer a billion dollar upgrade to substations, transmission and distribution systems (Lyons, 2019).

The adoption of NWAs holds the promise of reducing overall system costs and increasing the value of distributed assets. However, the identification of these projects may require the modification of traditional distribution-planning processes at many utilities (Lyons, 2019). It will also require fresh regulatory approaches, as there are embedded financial incentives under traditional regulation for investor-owned utilities to select capital intensive distribution projects rather than less-costly NWAs.

Integration of Electric Vehicles

"An internal combustion vehicle has one value: It drives people around. EVs are different. There are many value streams which can benefit third parties and customers. What V2G [vehicle to grid charging] does is liberate those value streams."

David Slutzky, Fermata Energy CEO, and founder (Walton, 2021)

At low levels of implementation, electric vehicles (EVs) have not yet had a significant impact on most utility distribution systems (Das, Rahman, Li, & Tan, 2020). However, with the transportation sector representing one of the largest sources of GHGs in most developed countries, policy makers are increasingly looking at EVs as a means of achieving GHG reduction targets and are implementing a range of regulations and incentives to encourage EV market penetration. At the end of the 2020, 13 countries and 31 additional regions have announced targets to phase out internal combustion vehicles, most by 2030 or 2040 (BloombergNEF, 2020), although it remains to be seen how many of these targets will be aspirational. Consumers are also being attracted to EVs by their dramatically declining costs, with the price of a lithium-ion automotive battery pack falling by about 90% between 2010 and 2020, from $1,110 per kWh to $137 per kWh (BloombergNEF, 2020). With continuing cost reductions arising from improvements in battery chemistry and manufacturing methods, it is forecast that by the mid-2020s, auto manufacturers will be able to build and sell electric versions of many models of mass-market vehicle as cheaply as those that operate with an internal combustion engine (BloombergNEF, 2020).

From only 17,000 vehicles on the road in 2010, by 2019 there were 7.2 million EVs world-wide, and while 47% of these were vehicles in China, there were 9 countries with more than 100,000 EVs on the road (IEA, 2020) (see Figure 8.4). The International Energy Agency forecasts that under existing government policies there will be roughly 140 million vehicles on the road by 2030, and under a scenario of more active government policy to achieve a 30% market share for EVs by 2030, there would be 245 million on the road (IEA, 2020). In terms of shaping the value chain of the typical distribution utility over the next few decades, if these forecasts come to pass, the impact of EVs will be substantial.

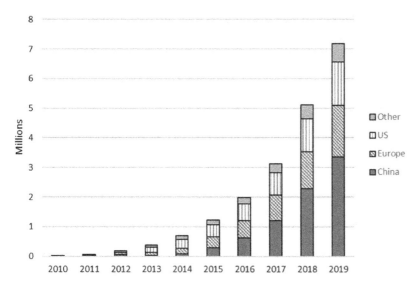

Figure 8.4: Global EV Stock, 2010–2019.
Source: Data from Global EV Outlook 2020, International Energy Agency (2020). EV stock includes both battery-only EVs and plug-in hybrid EVs.

The Challenge

Increased adoption of EVs will add increasing strain to many distribution grids. In the absence of controls or customer incentives, the addition of EV charging may greatly increase load on typical local distribution systems at particular times of the day. For example, the electricity consumption of a typical household in the United States is about 10,600 kWh of electricity per year (US EIA, 2020c), or about 28 kWh per day. A typical electric vehicle, such as a Nissan Leaf fitted with a 40 kWh battery, can charge in about 7.5 hours if attached to a standard "level 2" 240 volt charger that may be installed in a utility customer's home (Das, Rahman, Li, & Tan, 2020). If the utility customer comes home from work at 5 pm and plugs in their EV with a battery that is three-quarters discharged, it would add about 3 to 4 kw of demand to the local distribution system at a time that is already at peak evening consumption, and will consume as much electricity over that 7.5 hour period as would a typical household in a full day. This is probably manageable on most distribution circuits as the first few houses on a street install level 2 chargers for their new EVs. However, with this type of uncontrolled charging, as the number of EVs grows in a neighborhood, many of the local transformers and distribution circuits that service that neighborhood will typically not be sized to accommodate these increases in load. This was indicated by a 2019 study of German grids that found "In the future, actions by grid operators are necessary, as grid limits can not be fulfilled for high penetrations of EVs" (Held, Märtz, Krohn, Wirth, Zimmerlin, Suriyah, . . . & Fichtner, 2019, p. 18). This impact on the distribution system will also be dynamic, with EV-related load affecting different geographic areas at different times of the day or week, as drivers charge their vehicles at work, or at destinations like sporting events or shopping malls on the weekend (Ramchurn et al., 2012). In addition, the impact on electrical distribution systems will be amplified in areas with a concentration of EV charging, such as city bus stations, vehicle fleet depots or commercial truck stops (Das et al., 2020).

Mitigation and Opportunity

Although the growth of EVs may strain a local distribution network, this impact can be mitigated through control and management of the interaction between the vehicle and the grid. There are two main groups of tools that will manage this impact: policy and regulation, and grid controls. The first group of tools, policy, and regulation, can be used to encourage vehicle charging in a manner that will reduce the impact on the grid. This could include policies to encourage charging at work during daytime hours, when the loads on many electrical grids may be relatively low. It could also include time-of-use pricing of electricity, to incent customers to charge during off-peak periods. The financial incentive of off-peak rates can be substantial, with discounts in jurisdictions offering time-of-use pricing ranging from 5% to 40% in the United States (IEA, 2020), 20% to 56% in the UK (Zap Map, 2019 as cited in IEA, 2020), and 35% to 80% in France (IEA DSM, 2017 as cited in IEA, 2020).

The second group of tools to manage the interaction between the grid and the EV battery will build upon the "smart grid" digital control and communication technologies that many utilities have developed over the past decade. Grid operators and innovative suppliers are developing sophisticated scheduling and control systems to enable active management of the interface between the grid and EV batteries, although the implementation of these capabilities will require additional investment in system infrastructure (Das et al., 2020). Utilities in areas with relatively high penetrations of EVs are already implementing tools to allow the grid operator to manage the timing and rate at which an electric vehicle is charged on the grid (Donadee, Shaw, Garnett, Cutter, & Min, 2019). As the development of these control systems matures, the systems may support more complex algorithms, to balance factors such as the customer's travel schedules, load factors on the local distribution system, system-wide factors such as mid-day renewable energy over-production, financial incentives, variable electricity prices available to the customer, and compensation to the EV owner for deferral of charging (Ramchurn et al., 2012). Moreover, the system developed to balance these factors should ideally include learning algorithms to accommodate changes in electrical system loads (e.g., weekend vs weekday, winter vs summer), customer incentives, vehicle usage and other factors (Ramchurn et al., 2012). It will be a complex task, integrating measures of power grid system engineering, human behavior, and financial incentives.

These systems that manage the unidirectional flow of electricity from the grid to the EV battery are described as level one integration, or "V1G." A further level of tools being developed by grid operators, described as vehicle-to-grid, or "V2G," not only manages the charging of EV batteries, but also offers capability to utilize the EV battery as a potential source of electricity to the customer's home or as a resource to the grid. Under this level of control, EV batteries could be used as a grid resource to smooth system-wide supply and demand, or act as source of curtailable load in times of peak demand. EV batteries could also provide grid operators with services such as frequency and voltage regulation, or stabilization of isolated electrical networks (Das et al., 2020). Grid operators may shift EV charging schedules to accommodate a mid-day oversupply of renewable electricity or shift to later in the evening when other loads typically decline. Intelligent systems that can enable these capabilities, while minimizing negative impact to customers on the grid, will be a key element in integrating EVs into the operation of the electrical grid:

> so that now we can use all these devices and technologies, we can say "Hey, can we borrow some your storage? We will pay you." Or "Can we increase your load? We will charge your battery, because we have too much on the system, we have too much solar. And we will pay to charge your battery, or it will be free." You know what I mean?
>
> Utility executive, author's research session G1

While these V2G capabilities have seen implementation in some places in Europe and Asia, North American implementation has been limited to pilot studies at a few utilities due to technical constraints, safety issues, and the absence of a proven business case to justify the investment in the required systems (Walton, 2021). Nevertheless, if the forecast growth in market penetration of EVs over the next one or two decades proves to be real rather than hype, then the electrification of transportation will clearly be a key factor in the evolution of the utility value chain in the coming decades. The growth of EVs presents a tremendous challenge to the management of the grid, potentially requiring significant investment in distribution assets. Interestingly, the impact of the growth of EVs on aggregate demand for grid-supplied electricity may not be that great, with even relatively high levels of EV penetration resulting in only a marginal impact on generation and transmission assets (Muratori, 2018). By one recent industry forecast, EVs are forecast to account for only between two and 3% of global consumption of electricity by 2030, well within the capacity of planned infrastructure investments (BloombergNEF, 2020). In fact, in the United States and Europe this growth in electricity consumption by EVs will be largely offset by forecast consumption reductions arising from increased energy efficiency (BloombergNEF, 2020)[6]. However,

---

6 See Chapter 7 for further discussion of the variability of aggregate demand forecasts.

the impact on local distribution systems could be much greater, with clustering of EV deployment combined with uncontrolled charging routines resulting in increased peak demands on local distribution systems, requiring upgrades to system infrastructure (Muratori, 2018).

Fortunately, it is not all bad news for grid operators and owners. With the use of intelligent V1G systems and controlled charging, the required investment in local distribution systems to manage the impact of EVs can be mitigated. Furthermore, the integration of EVs into the grid using V2G systems offers grid operators the potential to use EV batteries as a new grid resource, a valuable asset in an environment with increasing levels of variable renewables-based energy. Under either level of system development, this alteration of the utility value chain will require rethinking other elements of the business model, including, for example, the firm's value offering (e.g., How does the utility define this new service that it is offering the EV owner?), customer identification (e.g., Will the utility interface directly with EV owners, or act through an aggregator?), and value capture mechanisms (e.g., How is the utility compensated for these services? Are the services priced in a regulated market, or unregulated?).

Value of Integrating Distributed Resources in the Value Chain

The value chain of the distribution utility has been developed over decades to efficiently deliver power to customers using a one-way flow of power. This remarkable architecture, that enables customers to change their level of consumption at any time, with assurance of quality and availability and at reasonable cost (Fox-Penner, 2014), has formed the foundation for a successful customer experience for many years. So, with all the inherent problems in intermittency and dispersion that distributed resources bring, why should these resources be integrated into this traditional value chain? Or, given the challenges of integration, would it just make sense to allow DER to develop separately, with restricted integration to the traditional grid? Should owners of distributed energy resources just be encouraged to go off-grid?

The answer is that the traditional grid and the newer sources of distributed variable energy resources will generally have greater value when integrated as a system than when operated as separate environments. There is an inherent value of integrating these two different environments that arises from improved system-wide efficiency and resiliency, and smoother supply and demand profiles. This inherent value comes from two primary factors: the connection of multiple sources of demand, and the connection of multiple sources of supply.

Connecting Multiple Sources of Demand

The history of the utility business model examined in earlier chapters demonstrated that when a utility can attract customers with varied demand profiles, it enables operation with increased system efficiencies and lower overall costs. Thomas Edison's Pearl Street station did not start to achieve financial profitability until the addition of customers running daytime electric motors to an original customer base that primarily used late afternoon lighting (Cunningham, 2015). The value of combining customers with differing demand profiles was also exploited by Sam Insull at Chicago Edison, adding factories and trolley companies to a customer base previously dominated by lighting customers (Wasik, 2006). Without combining customer groups with different consumption profiles, Edison and Insull would not have achieved the system efficiencies that they did. Similar opportunities are found in today's environment of distributed resources. If the surplus electricity from a rooftop solar panel at mid-day or a wind turbine at night can not utilize the grid to find a consumer or a means of storage, then the electricity it generates will be wasted. The grid allows distributed energy resources to connect with consumers with different demand profiles, enabling the more efficient asset utilization and the greater system-wide efficiency that comes with smoothing of demand (Fox-Penner, 2014).

Connecting Multiple Sources of Supply

In a world with a diverse portfolio of centralized and decentralized power resources, the grid provides an opportunity to interconnect these sources to operate most efficiently. A grid with a large volume of connected DER will enable an operating authority (which may or may not be the utility) to manage the range of supply sources as a portfolio, to deliver the least cost resources to users most efficiently (Fox-Penner, 2014). Also, due to intermittency of supply, most DER users require some form of backup. Certainly, if the grid is already available, it is more efficient to share backup capabilities through the grid rather than to have each generator developing their own system of backup (Fox-Penner, 2014). Finally, multiple sources of DER can add to system resiliency, reducing widespread customer outages during extreme weather events (NYPSC, 2015).

As an example of the potential of DER as a source of system resilience, consider the recent experiences in the state of California.[7] Following a devastating season of

---

7 California is often a source for examples, since its industry and regulatory response to emerging issues affecting electric utilities often lead other jurisdictions, both within the United States and internationally (e.g., industry deregulation, integration of renewables, etc.).

wildfires in 2018, California-based utilities implemented standard operating proce-dures[8] to de-energize transmission and distribution lines during periods of heat and wind to avoid dangerous wildfire risk (Morris, 2019). In the autumn of 2020, for ex-ample, as the risk of wildfires retuned to California, hundreds of thousands of utility customers faced rolling blackouts, as the vulnerability of traditional distribution was revealed (Romero, 2020). The great majority of customers in California do not yet have the capability to mitigate the impact of these increased outages by instal-ling their own distributed energy resources, or by linking with micro-grids powered by distributed resources (St. John, 2020). However, the state's distributed resource capacity is forecast to grow substantially over the next five years, from 5 gigawatts (GW) in 2020 to 13.5 GW (or about 30% of projected peak demand) by 2025, most of which will be in the form of EV charging and distributed batteries (Wood Macken-zie, 2020 as cited in St. John, 2020). If utilities and their suppliers can develop the technological and operational capability to manage those batteries as grid assets, and reach a regulatory agreement with the customers who own the distributed as-sets to make them available for grid requirements, the impact of rolling blackouts could be greatly mitigated (St. John, 2020). The experience and opportunities in Cal-ifornia will be applicable to jurisdictions anywhere dealing with increasing levels of distributed energy resources (including EVs). There will still be limitations in the batteries' ability to replace the grid, as EV batteries have unpredictable capacity when called upon, and behind-the-meter batteries are not yet commonly enabled by regulation to discharge to the grid (St. John, 2020). Nevertheless, with the develop-ment of operational capabilities and regulatory structures over the next few years, distributed battery resources could have the potential to be included as a significant source of supply in the management of the grid.

Summary

As the unit cost of electricity generated by distributed energy resources has tumbled steadily since the 1970s, this electricity supply has become increasingly cost competi-tive with traditionally supplied electricity. With these lower costs, utilities in many jurisdictions for the first time face the threat of competition from a competing source of electricity. As we have seen in this chapter, there are significant challenges to the

---

8 These operating procedures were developed in response to California's 150 thousand acre "Camp Fire" in 2018, which was started by sparks from the electrical grid of a utility. The utility pled guilty to eighty-four felony counts of involuntary manslaughter arising from the fire, which was blamed on a faulty power line (Associated Press, 2020).

integration of these resources into the grid (see Table 8.1). Nevertheless, the growth of distributed resources provides utilities an opportunity to integrate a significant source of new renewable supply to the grid. In addition, utilities will learn to utilize these energy resources as "non-wires" solutions to alleviate constraints on the grid, instead of relying on costly investment in traditional distribution hardware.

Table 8.1: Key Issues Reshaping the Utility Value Chain.

Key Issue	Description
Complexity of Integrating DER in the Value Chain	As DER achieves greater penetration, managing that source of supply is challenging within the technical framework of traditional distribution utility. New investment in systems and capabilities will be required.
Intermittency of Supply	As distributed renewable resources reach increased levels of penetration, the management of the grid increases in complexity. To manage this complex environment, it may become necessary for the utility to develop capabilities beyond its traditional boundary of the customer meter.
Dispersion of Resource	The one-way system simply has not been designed to accommodate significant volumes of two-way flow of electricity. The two-way flow of significant amounts of electricity will require investment in new hardware, systems, and procedures. But dispersed resources also deliver opportunities to address constraints in the grid, as a "non-wires" alternative.
Integration of EVs	Increased adoption of EVs will add increasing strain to many distribution grids, although with less impact on generation and transmission systems. It will be possible to mitigate this impact through control and management of the interaction between vehicle and grid.
Value of Integrating DER in the Value Chain	The grid inherently adds value in a DER environment from two factors: the connection of multiple sources of supply and multiple sources of demand. Connection of the two environments can enable improved system-wide efficiency and resiliency.

9 The New Business Model: Redefining the Customer

> I think the biggest mistake that the utility world has made in the first hundred years of its existence is to refer to our customer as 'the load.' To regard [the customer] as a passive device that sits at the end of the line.   Utility executive, author's research session F3

The "Customer Identification" element of a firm's business model defines the "targeted user and customer group" (Baden-Fuller & Mangematin, 2013, p. 421) around which the firm focuses its capabilities for product delivery (as discussed in Chapter 2). In a competitive marketplace, firms target new customers for expansion of their customer base, and analyze existing customers to identify unprofitable market segments, potentially leading to an exit from that portion of the market. Indeed, in the early days of the industry, this practice was certainly followed by investor-owned electrical utilities, with price competition to attract new customers in attractive markets, while less profitable customer segments (e.g., rural areas) remained underserved. However, as regulators and utilities in the 1920s and 1930s developed the "the regulatory compact," utilities gained status as monopoly providers with an "obligation to serve" all customers within their jurisdictions. Utilities could no longer use market segmentation to target profitable customers groups, and they could not offer differentiated products to certain customer groups (Smith, 1996). Utilities and regulators focused on the development of a cost-efficient value chain, developing a value offering characterized by quality and reliability. As noted in the quotation at the start of this section, the customer was often (and still is) referred to as "the load," a passive recipient of electrical service.

As much as utilities may have taken a minimalist approach to customer interactions, to an extent this inattention was mutual, with the customer having little interest in engaging with the utility. The nature of this relationship between utility and customer was confirmed in this interview:

> we would ask them [the customer] "Who is your electric utility?" And half of them would not know. Now, is that a problem, or is that okay? I think it is okay. It just goes to show you that electricity is about the light switch. It's about on and off. For the most part, if we are out of sight, we are out of mind.   Utility executive, author's research session C1

Despite this history of a minimalist relationship, utilities and customers in recent years have been developing new relationships where the customer is not simply a passive recipient of electrical service. There are several reasons for this relationship reset, including:

- **Capability:** With the investment of billions of dollars in smart grid communication infrastructure over the past decade (IEA, 2020), utilities have a much greater capability to communicate and interact with customers, and to offer new services to those customers.

https://doi.org/10.1515/9783110714036-012

- **Mutual Self-Interest:** Both customer and utility can increasingly benefit from interaction whether it is giving the customer access to new utility services (e.g., time of use rates), or giving the utility additional tools to manage the grid (e.g., enabling the utility to access customer air conditioning as a source of curtailable load).
- **Society's Interest:** It is in society's interest that utilities and customers find new ways to communicate and work together. Utilities receive mandates from policy makers to support key social policy objectives, such as the enablement of GHG reduction in the economy. Increasingly, the effective implementation of many of these policy initiatives requires interaction and communication between customer and utility (e.g., utility control of the timing and magnitude of EV charging).
- **Survival:** Finally, for the health of their business, utilities are facing an imperative to engage with the customer that has been long regarded as passive "load." For the past century, alternatives to grid-supplied electricity, such as customer self-generation, have been expensive enough that they were seldom a competitive threat. The utility's relationship was with a customer that had no practical, economic supply alternatives. However, with steep declines in the cost of distributed energy resources, utilities in many jurisdictions are at risk of losing their envied position as a monopoly service provider of electrical services.

The competitive threat of DERs as an alternative to grid-supplied electricity will vary between jurisdictions. For example, utilities in the Pacific Northwest of North America are blessed with relatively low-cost hydroelectric power and a cloudy climate, making rooftop solar less attractive as a replacement to grid-supplied electricity. Hawaii, on the other hand, with a legacy generation system dependent upon high cost fossil-fuels, has seen one of the world's highest deployment rate of rooftop solar (the share of single-family homes with rooftop solar is 33% on Oahu, 20% on Hawaii Island, and 27% on Maui) (Hawaiian Electric, 2020). Recent studies have found that the grid's cost advantage is disappearing across many jurisdictions, with distributed energy resources approaching "grid parity" (i.e., cost equivalence with grid-supplied electricity). For example, a 2015 North Carolina State University study concluded that for 93% of the customers in 42 of 50 of America's largest cities, electricity from a rooftop solar installation would be cheaper than the electricity purchased from their local utility (Kennerly & Proudlove, 2015). There are arguments that the low costs calculated in the study are achievable only due to government subsidies, and that the comparison does not reflect the value of firm electricity received from the grid (Potts, 2015). Nevertheless, it is also true that the costs of distributed energy sources have continued to decline since that study, and are forecast to continue to do so. The market penetration of alternatives to the grid are likely to continue to grow, supported, by policies such as California's first-in-the-nation building code that requires all new homes built after 2020 to have a solar photovoltaic system (Mulkern, 2021). These new building codes do not require the solar panels to be large enough to be the home's sole source of electricity, but clearly, California utilities are effectively no longer monopoly suppliers of electricity for many of their customers. Here and elsewhere, the utility's traditional relationship with the customer is certainly changing.

The Customer as Prosumer

In jurisdictions with increasing penetration of distributed energy resources, one of the most significant recent factors affecting a utility's relationship with the customer has been the rise of the "prosumer," a term developed by Alvin Toffler and first used in his 1980 book, *The Third Wave* (Toffler, 1980). Toffler used the term to describe the development of a future marketplace where customers not only consume a product, but also participate in the production of that product for their own consumption (Kotler, 1986). This term has been expanded beyond Toffler's original concept by electric utilities to describe consumers who not only generate some of their own electricity, but also sell some of that product back into the marketplace (e.g., customers with rooftop solar, selling surplus generation back into the grid). Under this relationship, the utility's customer can become a supplier. And, in fact, can be both supplier and customer at the same time. For utilities that have been accustomed to a monopoly relationship with a relatively passive customer, this is a disruption not only to their customer identification, but also to other elements of the business model, including value capture and the value chain.

> So, this is, in fact, a conundrum. And the reason that it is a conundrum is that regulatory fiats have forced a supply chain relationship [with the customer]. This is not a natural relationship. This is not like a match made in heaven, this is a match made at gunpoint. [Laughing] It is a forced marriage. Utility executive, author's research session G3

This emergence of the prosumer will have a profound impact on the customer relationship with the utility. For decades, the typical utility customer could be reasonably described as "the best customer in the world." This customer consumed the product without creating much fuss, paid bills on time, and showed up without fail year after year or, in fact, decade after decade. However, as this prosumer relationship has developed, this same customer has, in many respects, turned into one of the "worst customers in the world." As this utility executive recounted:

> it's not an apples-to-apples example, but a simple illustration. I happen to be a Starbucks fan. I've been going to Starbucks for the past 10 years and suddenly, a local city ordinance requires my local Starbucks to allow me [the customer] to take my surplus product and put it to them, as you will. So, every morning, instead of me driving in, and taking a latte from them . . . but now, I am brewing two [lattes] at home, consuming one, bringing the surplus into the Starbucks, putting it on their counter and telling them that this is theirs to deal with. [Starbucks] can do whatever they want with it; they can resell it right back to the customer behind me. But I am going to come back in the evening and take one from [Starbucks] and we are even. And by the way, while I am at it, I am actually going to use your restroom, and your Wi-Fi, and your

sweetener and creamer, and read your newspapers, and use your furniture and do some work here. So, that's where a lot of the utility prosumer relationships happen to be today. That particular Starbucks would not necessarily agree to do that, but it is being forced to by the local ordinance. Utility executive, author's research session G3

If "Starbucks" is replaced with "distribution utility" and "latte" with "electricity," it allows an interesting perspective on the relationship between utilities and customers that sell surplus electricity from a distributed resource into the grid. When only a few customers decide to pursue this prosumer relationship, then the coffee shop can cover its additional costs and lost revenue by increasing the price of everyone else's java by a few cents. However, if the number of prosumer-type customers continues to grow, then at some point the existing business model will no longer be sustainable. This will be as true for distribution utilities as it is for coffee shops.

The Customer as Competitor

Access to distributed energy resources is getting cheaper and easier. . .. And as that increases, the regulatory environment is going to become more complex. And this is where you're going to have scenarios where someone might say "well, I control the access point to the utility, beyond this point these are my customers." So, there is a lot of complexity in questions around this. I don't think anyone has really had to wrestle with that yet.

Utility executive, author's research session H5

Traditionally, the utility in a regulated environment has maintained a direct relationship with the end-consumer of electricity without a third party in the middle. (In a deregulated jurisdiction, much of the relationship with the customer may be taken up by a separate electricity retailer.) With the growth of DER technology, however, some utilities are starting to encounter a third party, such as a property developer, that installs its own DER resource, acting as middleman between the distribution utility and the end consumer. Under this new form of relationship, this middleman may become a competitor for those customers, operating and suppling the community with its own DER electricity, while utilizing the utility only for backup and certain services.

As an example, you can, as a developer, put up a condo unit, with individual civic addresses for each of the units, and you can contract with the public utility for one meter, a bulk meter. And then, the metering beyond that is managed by a third party. Not the utility. Now, in our environment, that's a real concern to us for a number of reasons. The primary reason is that, as soon as you have that situation, you are now having a new entrant into the market who is not necessarily playing by market rules. Utility executive, author's research session B5

This potential relationship, with another party as an intermediary between utility and consumer, will require regulatory direction which does not yet appear to exist in many jurisdictions. This is a new form of customer relationship, which places the utilities' value offerings and value capture mechanisms in flux.

Interacting with the Customer Beyond the Meter

Smart grid technologies and communications are changing the capability of the distribution utility to interact with the customer. With this enhanced capability, utilities must now choose whether they wish to have a role in operating or monitoring distributed resources installed on customer premises. There are at least two drivers for the utility to extend operational control to the customer side of the meter. First, to use the customer's resources to optimize the grid, and second, to allow the utility to provide enhanced services to its customers.

Using the Customer to Optimize the Grid

Traditionally, utilities have owned and operated assets up to the customer meter, but no further. This has been a boundary that utilities have been loath to cross. However, in an environment with growing numbers of customers with distributed resources on the customer side of the meter, the utility can better optimize the electrical grid as a system if it can reach across the customer meter, for example, to control EV charging, or control the supply of electricity coming from DER sources like rooftop solar. It is for this reason that many jurisdictions have specified that grid-connected distributed resources be controllable by the utility through devices like a connected "smart inverter[1]" that can be used by the utility to support grid functions or, if necessary, to curtail the resource. For example, Germany failed to require such control when installing their first wave of distributed solar resources and suffered such system instability that it was required to spend US$300 million to retrofit smart inverters into 315,000 existing solar installations to enable this control (Metcalfe, 2016).

> The utilities need control, visibility and control of these assets. And that's what Germany found out. And essentially what you want to do with the smart inverter is three things. Number one, you want to be able to control the output. Number two, you need to be able to control the voltage, to add vars or subtract vars to control the voltage. Number three, you need some kind of frequency response. Because what they were finding in Germany is, and you can imagine, if you have all this penetration of solar, and it is designed to disconnect as soon as it loses utility supply, you

---

1 See the glossary for description of "smart inverter.".

can have a frequency deviation that, without a smart inverter, causes all the solar to trip off. Now, you've lost all this solar generation, and it just compounds.

Utility executive, author's research session B5

Delivery of New Services to the Customer

The second factor supporting utilities' increased control of devices on the customer side of the meter, is the opportunity to deliver new information and services to the customer. Increasingly, capabilities are being developed to communicate inside the customer's premise to provide benefit to the individual customer or to the broader customer base.

> So, even if you set aside self-generation, even for those customers that are not necessarily candidates or interested or don't have the money to pursue that, all of our other customers are expecting more personalized services in terms of "help me manage my power."
>
> Utility executive, author's research session A5.

As an example of such as a service, a customer might allow some of their appliances, such as air conditioners, to be briefly turned off at times of peak demand, in exchange for compensation to the customer. These new services will require communications with the customer, but it is not yet clear who should manage these communications. In some early situations, the utility has invested in systems to deliver timely information to customers regarding their energy consumption and service options. However, some third-party companies, like Google, Amazon, and Tesla, are seeking business opportunities to position themselves to use energy information to strengthen their own customer relationship. Google's US$3.2 billion purchase of thermostat maker, Nest, is just one example of the interest shown in this area by some very large organizations (Wolsen, 2014). If the utility chooses to maintain the customer relationship itself, it will generally have to substantially deepen its own capabilities (Fox-Penner, 2014). Alternatively, if the utility chooses to work with partners that interface with the customer, such as Google/Nest who might position as an aggregator of energy consumption information and services, then the business-to-business relationship also requires a completely different set of capabilities and resources (Fox-Penner, 2014). Either way, the utility will be required to invest resources on the redefinition of this element of its business model.

> Customers are almost expecting that we may start to play, a little bit beyond the meter. They aren't necessarily expecting that we are going to build all of the in-house services and solutions ourselves. But potentially, together with an eco-system of service providers we will build those types of services to help them better manage their consumption.
>
> Utility executive, author's research session A5

Summary

It's not the technology, it is often the management of the customer that can be really difficult.

Utility executive, author's research session C1

Just as with other elements of the utility business model, there are emerging issues in the utility's customer identification, and the means it uses to communicate with those customers (see Table 9.1). It does appear that the utility's relationship with the customer will change as the demand for alternative services increases, although there are divergent viewpoints on the future shape of that relationship. Some in the industry see a much closer relationship than today, while others see a modification of the existing relationship only for those customers with distributed resources interacting with the grid. Others see the utility as a future member of a larger ecosystem that manages the customer's energy consumption, while others foresee the utility only as a "wires" company, with customer interaction managed by others. The future is tough to predict, although it does appear that future relationships will be different than the past.

Table 9.1: Key Issues Reshaping Customer Identification.

Issue	Description
The Customer as Prosumer	Regulatory change has forced a relationship where the utility's customer can become a supplier. And, in fact, can be both supplier and customer at the same time.
The Customer as Competitor	The middleman may become a competitor for distribution customers, operating and suppling the community with its own DER resources, while utilizing the utility only for backup and certain services.
Interacting with the Customer Beyond the Meter	*Using the Customer to Optimize the Grid* The utility can better optimize the electrical grid as a whole if it can reach across the customer meter to control the supply of electricity coming from DER sources. *Communication to Deliver New Information and Services* Technological capabilities are available to communicate inside the customer's premise to change the customer's usage, thereby providing benefit to the customer.
The DER Customer of the Future	Research indicates that the DER customer is going to be more challenging to service and will require more attention than a traditional utility customer. However, there are divergent viewpoints on the future shape of the relationship with the customers.

10 The New Business Model: Struggling with Value Capture

The value capture element of any business model performs an essential litmus test of the strength of that business model. In the long run, any commercial business must generate sufficient revenues to appropriately compensate employees, bondholders, and equity investors. If the value capture mechanism of the firm's business model can not be structured to meet those requirements, then the business will ultimately fail.

For most of the twentieth century, utilities' value capture mechanisms have counted among the economy's most robust. Shielded from competitors by a legislated monopoly, resistant to economic cycles with a product that is demand inelastic, and given access to a regulatory framework that allows assured recovery of costs and a return on capital, utility stocks and bonds have long been regarded as a boring but safe investment (Gilliland & Teufel, 2011). In fact, the value capture of the utility business model was so reliable that there were no large public utilities in the United States to declare insolvency in over fifty years following the Great Depression, until the 1988 bankruptcy of Public Service Company of New Hampshire, the state's major electric utility (Berry, 1988). The utility was forced to seek bankruptcy protection due to its inability to recover $2.1 billion it had invested in an over-budget nuclear power plant, deemed by the regulator to be imprudent (Berry, 1988). Since that time, and with deregulation of the generation sector in many jurisdictions, there have been more frequent insolvencies, particularly for firms exposed to the generation sector. The corporate failures have been due to a number of factors, including crippling losses in the early days of deregulation (e.g., PG&E in 2001) (Becker & Polson, 2005), a surplus of debt for new plant construction in the deregulated market of mid-2000s (e.g., Calpine in 2005) (Associated Press, 2004), and even some instances of outright fraud (e.g., Enron in 2001) (Bloomberg, 2001).

Nevertheless, the utility sector has continued to be valued by defensive investors for its low risk, low reward character (Gilliland & Teufel, 2011). These staid but predictable returns were noted by the investor, Warren Buffett, who in a 2020 interview described investments in the in the utility industry as "not a way to get real rich, but it is a way to stay real rich" (Yahoo Finance, 2020). However, Buffett also acknowledged in one of his renowned letters to shareholders that regulated utilities are facing a world of increasing competition from other sources, particularly renewables (Berkshire Hathaway, 2015). Although Mr. Buffett identifies government subsidies of renewable sources, rather than their sharp decline in cost over the past decade (see Chapter 7) as the source of this competition, he particularly identifies the growth of generation from renewables as a potentially adverse factor in the economics of incumbent utilities.

https://doi.org/10.1515/9783110714036-013

society has decided that federally-subsidized wind and solar generation is in our country's long-term interest. Federal tax credits are used to implement this policy, support that makes renewables price-competitive in certain geographies. Those tax credits, or other government-mandated help for renewables, may eventually erode the economics of the incumbent utility, particularly if it is a high-cost operator.   (Berkshire Hathaway, 2015, p. 23)

Whether the cause of growth of renewables is the availability of federal tax credits, or their remarkable decline in cost over the past decade, Mr. Buffett is shining the spotlight on changes in the utility value chain that are playing havoc with the value capture mechanism of some utilities. This is less the case with utilities that incorporate utility-scale renewables into their supply portfolio (e.g., large-scale wind and solar farms), as these fit more naturally into the traditional utility value chain of "generation-to-transmission-to-distribution-to-customer." However, the large-scale deployment of distributed electricity resources fits much less naturally in this traditional value chain and has the potential to erode the economics of the incumbent utility.

Value capture is a complex topic in a regulated industry, and library bookshelves are loaded with texts on rate design and cost of service regulation. However, this chapter has a narrower focus, and will concentrate on four areas particularly impacted by current changes in the utility business model:
- **Limits of Cost of Service Regulation:** Traditional regulation compensates utilities based on the amount of capital invested, and not based on the services used or the services provided.
- **The Paradox of Volumetric Rates:** The traditional rate structure can make it difficult for regulators to align the behavior of the utility with interests of the customer and society.
- **The Mismatch of Fixed Costs and Variable Rates:** Utilities typically recover a substantial portion of their fixed costs with volumetric rates. When a customer with distributed resources reduces consumption of grid-supplied electricity, then the utility will lose a portion of the revenue that would otherwise cover its fixed costs. Other customers will be required to meet this shortfall through increased rates.
- **The Value of Time and Place:** The traditional tariff paid by ratepayers applies uniformly to electricity consumed at any time and place on the system. However, the value of electricity generated by distributed resources can vary substantially by time and location.

Limits of Cost of Service Regulation

A Problem

Traditional regulation compensates utilities through "cost of service" regulation, with profits based on the amount of capital invested by the utility, and not based on the services used, or the services provided. How will distributed resources, which are often not owned by the utility, fit into this framework?

> if our role is to choose the lowest cost resource for customers, we want to make sure that we have considered DERs [distributed energy resources] as one of those resources, in those situations where it makes sense. And if so, how should they be considered in how we make money? Will that continue to erode our opportunity for rate-base? Or do we essentially become the network operator that owns some of the assets and contracts for others. And then, how do we make money off that? Utility executive, author's research session G1

One of the basic premises of the "cost of service" regulation developed in the 1920s and 1930s (which was described in Chapter 6) is that while a utility can recover its operating costs through its rates, its profits are determined by the regulated return on the utility's rate base[1] (i.e., how much has the company invested in utility assets to serve its utility customers) (Ralff-Douglas & Zafar, 2015). This mechanism has a built-in bias to encourage capital over-investment by utilities to increase their rate base (a bias known to economists as the as the Averch-Johnson effect [Averch & Johnson, 1962]). This effect has sometimes resulted in a joking observation that if a utility wishes to boost its return, it has only to invest in the renovation of the CEO's office. Nevertheless, when combined with regulatory oversight, it has been generally agreed that cost of service regulation has worked well to align the interests of utility shareholders and their customers through most of the latter twentieth century (Ralff-Douglas & Zafar, 2015). However, this regulatory framework does not work well in a value chain with grid assets, such as EV batteries and rooftop solar generation, that are increasingly owned by others. There are two fundamental reasons for this: first, the method of compensation for capital investments, and second, the method of compensation for operating costs.

First, if the capital investment required for a distributed energy resource, such as rooftop solar or a battery resource, is incurred by the customer or by a third party, as is typical, then that capital falls outside of the utility's rate base. Under the traditional cost of service model, the utility is left without a financial incentive to encourage investment in the distributed resource, or to manage the energy from the distributed resource once it is in place.

Second, while the operating costs of managing the distributed resource are an expense for which the utility can recover its costs under the traditional cost of service model, it is not a source of profit. Operating costs are ordinarily expensed for accounting purposes, do not contribute to the utility rate base, and hence, do not contribute to the utility's profit. It can certainly be argued that if the utility is providing services to the customer, then the utility should receive equitable compensation for that service.

---

1 See the glossary for definition of "rate base."

However, these objectives will be difficult to achieve using traditional "cost of service" regulation in an environment of growing DER.

A Potential Remedy

Cost of service regulation incents utilities to pursue alternatives that invest in capital assets rather than those that result in operational expenses. This approach is increasingly misaligned with the direction of today's economy, which is increasingly service based. Firms that once would spend tens of millions of dollars on data centers or on the implementation of enterprise software systems now outsource to cloud-based service providers, spending operating dollars instead of capital. Firms that in the past would have once built their own logistics system, with trucks and warehouses, now outsource to logistics specialists. Activities that once were supported by capital investments are now available at lower cost from third parties. However, under ordinary accounting rules, the purchase of these types of services from third parties are expensed, and not added to the rate base. Utilities are incented by cost of service regulation to invest in their own assets, rather than purchase services from third parties.

Utilities have a similar issue when faced with increasing penetration of distributed energy resources. Even though it may be in the public interest to allow customers to invest in their own electrical resources, under cost of service regulation utilities are incented to build and own generation capacity themselves (thereby earning a rate of return through the rate base). They have no financial incentive to manage distributed resources owned by someone else. The cost of managing such a resource could be recovered by the utility through its rates, but without profit margin for the utility.

Regulators are looking for new ways to modify cost of service regulation to create methods of aligning utilities' revenue streams with the public interest. One example is a performance-based regulatory model being developed by the UK Office of Gas and Electricity Markets (Ofgem) (Spiegel-Feld & Mandel, 2015). This framework, referred to as RIIO (Revenue = Incentives + Innovation + Outputs) is designed with the objective that utilities be indifferent whether an expenditure is classified as capital (and added to the rate base) or operating expenditure. It does so by including both types of expenditure in a single regulatory asset class which earns a predetermined rate of return. This, in combination with a revenue cap, is structured to give utilities an incentive to seek the most cost-effective solution for ratepayers (Spiegel-Feld & Mandel, 2015).

Another example is found in California, where new ways are being investigated to compensate utilities for undertaking services, rather than investing capital. The California Public Utilities Commission (CPUC) has undertaken pilot projects that would allow utilities to earn a 4% incentive fee on payments to owners of distributed energy resources (CPUC, 2017). Unlike traditional regulation, this will enable the utility to be compensated for the task of integrating and managing the DER resources into the distribution grid, even though it does not own the DER asset.

The Paradox of Volumetric Rates

A Problem

The traditional rate structure can make it difficult for regulators to align the behavior of the utility with interests of the customer and society.

> So, the current business model is inherently focused on a linear pipe model that is focused on throughput and the amount consumed at the end. That model is dead and gone, it died some time ago, and we just haven't buried it yet.   Utility executive, author's research session E3

The use of volumetric charging (i.e., a charge that is based upon the amount consumed) to match costs and revenues has been in use since the days of Thomas Edison charging customers of his Pearl Street station "by the light bulb" (Wasik, 2006). As metering technology improved, utilities refined their techniques of measuring consumption until they could charge customers based on metered electricity consumption "by the kilowatt hour," a practice that continues at the core of the utility revenue model a century later. Although bills at many utilities also include a monthly charge, most utility revenue is still collected through a charge dependent upon the volume consumed by the customer.

Volumetric rates were advantageous to both the electric utilities and their customers in the early years of the utility industry. Significant scale economies on the supply side of the industry, coupled with rising electricity consumption on the demand side, enabled a steady decline in rates, making for a relatively non-confrontational relationship between utilities and regulators for many years. Furthermore, volumetric rates are simple, easy for customers to understand, and by their very nature, encourage customers to be efficient in the consumption of electricity.

Today, however, utilities and regulators using a traditional volumetric rate structure will find that they are faced with a paradox. On one hand, it is in the interest of society for utilities to assist consumers in using less electricity, not only to reduce costs, but also to reduce the environmental impact of generating and transmitting the electricity. The social mandate of an electrical utility should support a customer's efforts to become more energy efficient, and to consume less electricity. However, a utility that is compensated through volumetric rates can improve its profitability through increased consumption of electricity. This inherent structure can make it difficult for regulators to align the behavior of the utility with interests of the customer and society.

A Potential Remedy

The rates a utility charges its customers are normally set every few years through a "rate case"[2] whereby the utility and regulator examine forecast costs and electricity consumption and agree a rate that will be effective for a period of years, until the next rate case.

> Unit price = (Forecast total annual cost of the utility's regulated operations)
> divided by (the forecast annual quantity of electricity sold)

However, once the rate is set, if electricity sales fall below those forecast in the rate case, ratepayers will under-pay their share of fixed costs, and the utility's earnings will be less than forecast. Conversely, if the utility can boost sales beyond those forecast, it can over-recover on fixed costs, ratepayers will overpay for their share of fixed costs, and the utility will boost its profits. This factor, referred to as a "throughput incentive," has been found to be an impediment to aggressive investment by utilities in energy efficiency (NAPEE, 2007).

Regulators have sought various mechanisms to eliminate this natural bias that utilities have against supporting customer energy efficiency. One of the regulatory tools developed to achieve this is known as "decoupling," which is designed to ensure that a utility is indifferent to the amount of electricity it sells. There are many different variations of this regulatory mechanism (NREL, 2009), but it most typically operates by automatically altering up or down the rate paid by the customer, to adjust for changes in sales levels. The impact on a customer's bill is typically small, with a less than 1% impact on most customers' bills (NREL, 2009). However, when multiplied over many customers, the impact can be significant enough to ensure that the utility's recovery of fixed costs stays at the levels agreed in the current rate case with regulators, including a fair return on investment. It also ensures that ratepayers contribute to their fair share of fixed costs, although individual customers can still reduce their share of costs by reducing their consumption (NREL, 2009).

The many variants of "decoupling" are just one of the tools that policy makers have available to increase the probability of success of energy efficiency policies. It has proven to be an effective tool for policy makers to use as part of a broader initiative to ensure that the interests of utilities and society are aligned. Decoupling of some variant has been implemented for electric or gas utilities by regulators in thirty American states, and is being studied in an additional twelve states (Lazar, Weston, & Shirley, 2016).

---

2 See the glossary for a description of "rate case."

The Mismatch of Fixed Costs and Variable Rates

A Problem

Utilities typically recover a substantial portion of their fixed costs with volumetric rates. When a customer with distributed resources reduces consumption of grid-supplied electricity, then the utility will lose a portion of the revenue that would otherwise cover its fixed costs. Other customers will be required to meet this shortfall through increased rates.

Net Metering

> And net metering is one of the worst things to happen in a lot of these jurisdictions, because they finally figured out that these guys [i.e., customers], by putting solar panels on the roof, weren't paying for the wires. And they are not just bypassing [the cost of generated] energy, which is fair, they are also bypassing [the cost of] wires, just completely unfair. And people, then, that don't get solar panels, are required to pay more. So, they are cost shifting onto people that don't have solar panels.   Utility executive, author's research session D1

Increasingly, customers are investing in the ability to generate or store their own electricity, and this can create financial challenges for utilities. Even if customers with distributed resources do not sell surplus production back into the grid, the amount of electricity that they purchase from the grid will decline, and the utility will see its volume driven revenue diminished. This creates a financial problem for the utility since it must still incur the fixed cost of maintaining capacity to supply that customer from the grid. This financial problem can be even greater if the grid receives surplus electricity from the customer's distributed generation, since once certain levels of generation from distributed resources have been attained, the utility must invest to upgrade the distribution system to support two-way flows of electricity.

To compensate customers with DER who sell their surplus energy back into the grid, regulators in many jurisdictions have developed a "net metering" tariff that offsets the electricity that a customer feeds into the grid against amounts consumed. In effect, the producer of electricity is compensated at the same tariff rate[3] as the grid-

---

3 In the United States, the average electricity rate is composed of the following costs: 54% for generation; 36% for the product delivery, including transmission, distribution, customer billing and contact, and programs such as demand side management (DSM), and 10% for other costs, including administration and general expenses (Aniti, 2017).

supplied electricity consumed. With the net metering rate set at this level, the utility is effectively refunding the customer for avoided variable costs, and for the unavoided fixed costs of the electrical system (Eisen, 2013).

How much cost can a utility typically avoid if its volumes decline? It is estimated that fixed costs represent between 40% and 65% of the electric utility bill of a typical residential customer in the United States (Wood, Hemphill, Howat, Cavanagh, Borenstein, Deason, . . ., & Schwartz, 2016). Accordingly, with an average residential electricity bill of about $115 per month (US EIA, 2020c), the monthly fixed cost of serving a residential customer is in the range of $46 to $74 per month (Wood et al., 2016). However, the amount of actual fixed charge on a customer's bill in the United States is about only about $10 per month (Wood et al., 2016), less than 10% of the total charge. Clearly, fixed charges to the customer at the average utility are not sufficient to compensate utilities for their fixed costs of operation, and these utilities are reliant on volume driven charges for recovery of fixed costs.

Volumetric Rates and Social Financial Equity

> But even there, solar is still a rich man's game. By that I mean, when we were in [a jurisdiction with rooftop solar customers], if you drive into a wealthy neighborhood, solar penetration is pretty high. Probably half the houses have some sort of solar arrangement. If you go into a middle class, maybe 20 to 30 percent of the homes. If you go to a poor neighborhood, it doesn't exist at all.
> Utility executive, author's research session C3

A universally accepted social objective for the setting of a utility's rates is to ensure that, while the utility collects sufficient revenue to cover its operating and capital requirements, the distribution of costs to ratepayers should be fair and equitable (Bird, McLaren, Heeter, Linvill, Shenot, Sedano, & Migden-Ostrander, 2013). In many jurisdictions, the equitable distribution of costs has become more difficult with the growing implementation of net metering. Under the net metering tariff mechanism, the producer of electricity from a distributed resource, such as rooftop solar panels, is often compensated at the same unit rate per kilowatt hour as the electricity consumed. However, as noted in the prior section, when a utility's customer generates their own electricity and reduces consumption of grid-supplied electricity, the only costs avoidable by the utility are variable costs. Since the utility can not fully defray fixed costs in this situation, and the utility is allowed to recover its costs, the fixed costs must be transferred to other users who do not have distributed resources. For example, a study of customers in California found that once they had installed rooftop solar, residential net metering customers reduced their contribution to the utility's fixed cost of service by 43% on average (CPUC, 2013). Instead, these costs are transferred to customers without rooftop solar generation. If unaddressed, this practice will tend to create inequity of cost sharing between customers that have distributed resources, such as rooftop solar and batteries, and those that do not. A further unfortunate factor is that the inequity of cost sharing often falls between high- and low-income households, with low-income households bearing a disproportionate share of the cost burden (CPUC, 2013).

A Potential Remedy

> Well, as you know, most of our costs are fixed. Almost all of our revenue source is variable, with a little tiny, fixed piece, for connection charge. And that is a recipe for a problem. [Laughing] So, our goal, although we won't be there quickly, will be to move toward rates that more closely reflect cost. Utility executive, author's research session C1

The composition of most utilities' rates is not reflective of the fixed and variable cost structure of the services they provide. This can create problems of sending incorrect price signals to consumers of grid-sourced electricity, by allowing customers with net metering to avoid their appropriate share of fixed costs and placing an unfair share of fixed costs on other users.

A solution sought by much of the industry would make charges more reflective of underlying fixed and variable costs, with larger fixed costs recovered through a flat charge, and variable costs recovered through volume-based charges. This would not only improve the price signals between supplier and customer, but also lessen the inequitable distribution of costs between customers who have distributed energy, and those who have not. However, attempts by several jurisdictions to implement increased fixed charges have met strong public resistance (Flessner, 2018). Opposition often arises from three constituencies (Wood et al., 2016). First, there is a public perception that the impact of an increased fixed charge will fall disproportionately on lower income groups, creating opposition from anti-poverty groups and politicians (Flessner, 2018). Second, it is argued that fixed charges are not as inherently effective as volume-based charges in sending price signals to consumers that encourage energy efficiency (Flessner, 2018). Third, companies selling distributed energy resources also object to a higher fixed utility charge, since it reduces the comparative financial benefit of their product. For example, when an Arizona utility sought to substantially increase the fixed charges on customers with grid-connected solar panels, they were sued by the state's largest vendor of rooftop solar installations, arguing "anticompetitive and tortious conduct designed to eliminate solar competition" (Randazzo, 2018, para. 3).

The Value of Time and Place

The Problem

The traditional tariff paid by ratepayers applies uniformly to electricity consumed at any time and place on the system. However, the value of electricity generated by distributed resources can vary substantially by time and location.

> we simply determine what we think the price for the purchase of energy from a solar rooftop system should be, and we post a tariff, and everybody gets the same price.. . . it fails to recognize the difference in cost and value across time, and in different parts of the system.
>
> Utility executive, author's research session E4

We all know the value of being in the right place, at the right time. Electricity on a distribution system works that way as well, as values can vary substantially by time and location. For example, distributed energy resources can have significant value if they help the utility to alleviate demand on overloaded grid assets, thereby deferring or eliminating the requirement to upgrade that distribution equipment. Conversely, distributed energy resources will have limited value on the grid if generated at a time of day with little demand, and on a circuit with no access to storage (e.g., mid-day in a residential neighborhood is often a time of low demand, but a time for peak electricity generation by residential rooftop solar). However, a rate paid for distributed electricity that is common across a jurisdiction will fail to reflect this geographic and time-related variance in value. A common rate will not provide utilities the opportunity to use pricing to change users' consumption patterns or to incentivize changes in energy production patterns from distributed resources. As discussed in Chapter 8, this issue will gain in importance if the penetration of EVs meets growth forecasts over the next decade. The addition of charging of EVs may greatly increase load on typical local distribution systems at specific times of the day.

A Potential Remedy

The introduction of smart metering and other smart grid technologies will provide utilities with the ability to differentiate charges by time and location to match prices more closely with marginal cost. One method available to utilities is time of use (TOU) pricing, which increases prices during periods of peak periods, and discounts rates at periods of low usage (e.g., to encourage charging of EVs at night). In this way, the pricing signals sent to consumers can better reflect the underlying costs of the system and encourage customers to modify their consumption to increase efficiency of the grid.

There are tremendous opportunities for growth of this tariff mechanism. Although almost 100 million smart meters have been deployed in the United States in 2020, only about 4% of these accounts had a time of use pricing scheme available to them (Faruqui & Bourbonnais, 2020). However, greater penetration has been attained in other counties, with Italy achieving 75% to 90% participation on a country wide basis, France 50%, and Spain 40% (Faruqui & Bourbonnais, 2020). Even within the United States, there is higher penetration in specific regions, including Michigan at 75% and Maryland at 80% (Faruqui & Bourbonnais, 2020).

The development of unique tariffs to differentiate by time and specific geographic regions are still in relatively early stages of adoption. There are significant hurdles to the inclusion of these tools in a firm's value-capture capabilities, requiring smart grid investment in IT systems, electrical distribution system, sensors, and controls. It also critically requires public education and customer engagement. However, as the penetration of distributed resources continue to grow, and as the charging of EVs gain greater penetration, the value of these alternative pricing schemes will increase, and the imperative to implement them will continue to grow.

Summary

Warren Buffett was correct. The increase in renewable resources has the potential to fundamentally alter the economics of the environment in which incumbent utilities operate. As the penetration of these resources increase, it will expose problems and risk (see Table 10.1) in a utility value capture mechanism that has produced decades of safe, solid returns to employees, bondholders, and equity investors.

Table 10.1: Key Issues Reshaping Value Capture.

Issue	Description
Cost of Service Regulation	Traditional regulation compensates utilities based on the amount of capital invested, and not based on the services used, or for the services provided.
	One solution, implemented in the UK, is designed such that utilities should be indifferent whether an expenditure is classified as capital or operating expenditure.
	Another solution, implemented in California, allows utilities to earn a return for the provision of services.
The Paradox of Volumetric Rates	The traditional rate structure can make it difficult for regulators to align the behavior of the utility with interests of the customer and society. Utilities are incented to increase sales, while customers and society seek lower consumption.
	One of the regulatory tools developed to address this is known as "decoupling," which is designed to ensure that a utility is indifferent to the amount of electricity it sells.

Table 10.1 (continued)

Issue	Description
The Mismatch of Fixed Costs and Variable Rates	A large share of the average utility's fixed cost is recovered through variable charges. This can create problems of sending incorrect price signals to consumers, and an inequitable sharing of fixed costs. A solution sought by much of the industry would make charges more reflective of underlying fixed and variable costs, with larger fixed costs recovered through a flat charge.
The Value of Time and Place	The traditional tariff paid by ratepayers applies uniformly to electricity consumed at any time and place on the system, although the value of this electricity can vary substantially by time and location. Utilities are developing the capabilities to differentiate charges by time and location to match price more closely with marginal cost.

11 The New Business Model: Redefining the Value Offering

The "Value Offering" element of a firm's business model defines the method by which "the organization engages with the customer to create value" (Baden-Fuller & Mangematin, 2013, p. 421). One famous example of a value offering was provided by Henry Ford, who is reputed to have said, "Any customer can have a car painted any color that he wants, so long as it is black" (Alizon, Shooter, & Simpson, 2009, p. 1). Ford's limitation of the Model T to a single color was not due to an unimaginative design sense, but rather it was driven by an unrelenting focus on process improvement. It is speculated that Ford had determined that black paint dries faster than other colors (Alizon et al., 2009), and faster drying paint was just one of the many factors that cut production time for a Model T from 12 hours to 93 minutes within four years (Brown & Anthony, 2011). In the case of the early years of the Model T, Ford's value offering to the customer was clearly focused on reliability and low cost, rather than style, which was totally consistent with his single minded focus on value chain efficiency.

Like the early Henry Ford, utilities and regulators focused on supply side factors after the regulatory compact was formed in the 1930s (Smith, 1996). The result was a value offering that focused on reliability and cost, but a product that was largely undifferentiated between firms (Smith, 1996). Customers did not have an ability to choose between product attributes, and every customer on the street received the same service from its utility. In fact, a person could move from one side of the country to the other and largely find the same electrical service from different utilities.

This focus on standardization across the industry was a departure from the earliest utility business models, where utilities offered different services to different customers, depending on their requirements. For example, Edison's early central stations ran electrical service best suited for lighting, at 110 volts, while other utilities offered higher voltages better suited for railcar companies or electric motors in factories (Bakke, 2016). But with the regulatory compact, product differentiation largely came to an end. Although the capabilities a utility needed to deliver reliable quality to the customer had many components, including energy, capacity, reactive power, and various reserves (Satchwell & Cappers, 2018), these elements were bundled into a single product offering, and the resultant service offered for many years to most customers was "any color that they want, so long as it is black."

In the 1980s utilities did start to deliver some additional services to customers, as regulators directed utilities to make demand side management[1] programs available to encourage their customers to become more energy efficient. In the 1990s and

---

[1] See the glossary for definition of "demand side management."

https://doi.org/10.1515/9783110714036-014

2000s, newly deregulated electricity retailers, many of them spun off from regulated utilities, offered varied product offerings to customers (Satchwell & Cappers, 2018), and some utilities started to offer "green" electricity, enabling customers to pay a small premium for electricity generated by a renewable energy source. In recent years, utilities in some jurisdictions have ventured into areas that are traditionally unregulated, such as financing or selling electric vehicle chargers and solar photovoltaic systems (Satchwell & Cappers, 2018).

Despite these limited forays into alternative or nonregulated products, the revenues of regulated electric utilities remain largely derived from traditional regulated sources. However, the declining cost of distributed resources and the forecast growth of EVs is driving a new demand for electric services, and the emergence of smart grid technologies has increased the capability of utilities to deliver those services. In this new environment, utilities are re-examining how they engage with the customer to create and deliver value.

This chapter will not deal with the traditional value offerings of the utility: the reliable delivery of low cost electricity to the customer's meter. Instead, the chapter will concentrate on key questions driven by changes in the utility business model, and particularly by a value chain that includes a growing presence of renewable and distributed resources.

- **The Supplier of Grid Services:** What inherent value does the grid offer a customer with distributed resources?
- **The Enabler of Clean Energy:** How do regulated utilities support their customers to become consumers of clean energy, whether from the grid, or from distributed resources?
- **The Position in the Ecosystem:** In the growing ecosystem of unregulated suppliers to an increasingly distributed grid, in what areas of that ecosystem should the utility compete?

The Supplier of Grid Services

Once a customer has invested in distributed resources, what value does the customer receive from their continued connection to the grid? In addition to the delivery of reliable electric power, there are fundamental characteristics of the grid that contribute ongoing value to customers with distributed resources: backup, frequency and voltage services, and interconnectivity.

The first inherent value of the grid to customers with distributed resources, is that it provides low cost, highly reliable backup. There will be times when distributed resources such as wind or solar are not available due to windless or overcast conditions, equipment failure, or regular maintenance. For customers who go totally off-grid, it is costly to maintain an assured supply of standby power through a backup generator and batteries. It is generally much cheaper to stay attached to the grid.

The second inherent value of the grid to customers with distributed resources is as a source of frequency and voltage control services, which are needed when converting DC electricity generated by solar or wind to AC power used within a house or business. Typically, the power inverter[2] of a distributed energy resource, such as rooftop PV installation, relies on the grid to obtain a valid voltage and frequency reference. If disconnected from the grid, or in the event of a grid outage, many distributed resources can not function without these grid-supplied references (Wood & Borlick, 2013). It is possible to operate distributed resources in the absence of the grid, but it requires further customer investment in a "grid-interactive inverter," batteries or generators, and often the installation of electrical panels and wiring to service the reduced loads in the building to be run by backup power (e.g., refrigeration, lighting) (Mirafzal & Adib, 2020). Most customers with distributed resources have not made the investment in equipment required to operate without the attachment to the grid, although areas with severe storms and severe weather patterns have seen an increase in customers doing so (Wood & Borlick, 2013).

The third inherent value of the grid is its "interconnectivity." That is, the grid interconnects distributed resources to a very controlled marketplace, providing an outlet to sell surplus electricity to other consumers, and to buy from the grid when required. Just as early innovators of the utility business model in the early 1900s discovered the efficiencies of connecting previous stand-alone electrical systems, the same is found today, as the grid provides a venue for the sale and purchase of surplus distributed energy output.

This value of connecting to the grid has some relationship with a proposition developed in the early days of the internet, known as "Metcalfe's Law." This proposition was based on the observation that as the number of telecommunication nodes in a network increase, the cost of adding additional nodes increases in an arithmetic progression, but the number and value of connections between the nodes increase as a square of the number of nodes (Zhang, Liu, & Xu, 2015). Intuitively this general relationship makes sense in a telecommunications network, as one telephone in a network is of no value, but two telephones is of some use, and three even more. This relationship has also been used to help describe the formidable valuations earned by some of the early internet titans like Facebook and Google (Zhang et al., 2015), and more recently of firms building out platform business models such as Uber and Airbnb. However, the strict application of Metcalfe's law in examining the utility sector would be a misapplication (Briscoe, Odlyzko, & Tilly, 2006). First, the law assumes an equal value of connections, which is not the case in an electrical distribution network (Briscoe et al., 2006) (two rooftop solar panels connected to a grid on opposite sides of the state are of limited value to the grid as a whole). Second, the law fails to recognize the cost of managing the complexity of electrical

---

2 See the glossary for definition of "inverter" and "grid interactive inverter."

networks with thousands of distributed nodes (Briscoe et al., 2006). Nevertheless, Metcalfe's law is a useful concept for understanding that a distributed resource has greater value when connected to other producers and consumers, than as a stand-alone asset. Just as early innovators of the utility business model in the early 1900s discovered, there are efficiencies of connecting previous stand-alone electrical systems. This same value is found today in connecting DER producers with other producers and consumers, to provide a venue for the sale and purchase of surplus electrical output.

The Enabler of Clean Energy

> The customers are very interested in that. They want to reduce their footprint, they want to reduce their greenhouse gases, and if we can show them that they are doing that by staying connected to the grid while using renewable resources as their source of energy, then I think we would be happy with that aspect of staying connected to the grid.
>
> Utility executive, author's research session D3

Because of the backup service that it provides, the connection to the grid enables customers who wish to install and consume renewable distributed energy to do so without the risk of frequent outages. This growing group of customers are strongly motivated to take steps that they perceive will reduce their environmental footprint. According to a recent iteration of a multi-year study (Motyka, Thomson, Hardin, & Sanborn, 2020), in 2020 respondents who had installed rooftop solar responded for the first time that their primary driver for doing so was that it "is clean and does not contribute to climate change" rather than cost saving (Motyka et al., 2020, p. 12).

The same study confirmed the value of "green energy" to many customers, finding that roughly one third of respondents would be willing to pay more to access it (Motyka et al., 2020). In addition, almost three quarters of business respondents indicated that their customers want them to procure a certain percentage of their energy from renewable sources (Motyka et al., 2020). Certainly, when the security of being attached to the grid enables a grid-connected ratepayer to consume renewables, there is an intangible value to the ratepayer.

In addition to enabling individual ratepayers to pursue environmental objectives through the installation of distributed resources, the grid is also emerging as a valued tool to support policy makers in addressing issues of GHG reduction. It does so in two ways. First, the grid can support the decarbonization of key sectors of the economy, such as transportation (as was discussed in Chapter 8). Second, the grid can support the implementation of renewable generation resources, both central and decentralized. In the United States, thirty states and three territories have adopted renewable portfolio standards (RPS) setting targets for clean energy (NCSL, 2020), most of which specifically include provisions for distributed renewable resources (Kassakian, 2011). In Washington, the administration is putting in place a federal framework to deliver

100% clean electrical energy in the United States by 2035 (Bussewitz & Krisher, 2021). In Canada, the federal government has rolled out a nation-wide carbon tax that will incent utilities to reduce their carbon footprint through increased generation using renewable resources (Curry & Giovannetti, 2018). As signatories to the Paris Agreement face looming targets for GHG reduction, countries across the world are setting legally binding targets to reach net-zero emissions by 2050, with significant new commitments from China, Japan, South Korea, Canada, United States, United Kingdom, and the European Union (Rathi, 2021). Many of these commitments target the electric sector as a significant source of that reduction. As noted by Moody's Investors Service:

> Because the electric grid is a critical piece of infrastructure that is a vehicle for policymakers to implement their energy policies . . . it will become even more important as the platform for the more complex flows of power and information in the utility of the future.
>
> (Tiernan, 2014, p. 1)

The Utility's Position in the Ecosystem

Traditional thinking of business strategy often starts with an examination of the positioning of a single firm in a competitive marketplace (Porter, 1979). However, management research in recent years pays greater focus on the value delivered by that company as a member of a larger value-creation ecosystem (Massa, Tucci, & Afuah, 2017). This perspective views a business model as an "ecosystem," extending past the individual firm's boundaries, and including interdependent activities of other firms operating as suppliers, partner firms, and even competition (Zott & Amit, 2010). A key issue faced by many firms is determining which elements of the ecosystem to join as competitive participants, and which should be left to partners and competitors (Massa et al., 2017). With the rise of distributed resources and the increasing complexity of the ecosystem delivering electrical service to customers, this issue has risen in importance for many utilities.

The ecosystem that delivers electricity in many jurisdictions has become much more complex in recent years. A major contributor to this complexity is the digitization of the value chain, which enables the utility to integrate its supply chain with other members of the ecosystem (e.g., a distributed battery or rooftop solar system owned by another party). A second contributor to this complexity is the growth of distributed energy resources, and the diversity of firms that deliver them. A recent exhaustive study (Pérez-Arriaga & Knittle, 2016) identified 144 different business models in use by firms engaged in the delivery of distributed energy to consumers. These firms operate in three main categories: demand response and energy management systems, electrical and thermal storage, and solar photovoltaics (Pérez-Arriaga & Knittle, 2016). These firms offer varied combinations of installation, ownership, financing, operation, and maintenance of distributed energy products.

The firms delivering demand response[3] services are carrying out an activity some-times performed by a utility itself, under the direction of its regulator (Pérez-Arriaga & Knittle, 2016). Other services, however, are supplied in competitive markets that utilities generally do not enter. These marketplaces do not bear the characteristics of a natural monopoly in which utilities traditionally operate, and regulators need to question whether a utility operating from a protected monopoly in one market, would be unfairly utilizing its resources to compete in another. To move into these markets as a competitor requires a fundamental change to the traditional boundaries of the utility. For example, there is a well established network of firms that will install, own or finance battery storage systems at the premise of commercial and industrial custom-ers, to enable these customers to avoid expensive charges at periods of peak demand (Pérez-Arriaga & Knittle, 2016). This service, which utilizes assets on the customer side of the electric meter, exists outside the usual boundaries of services offered outside a traditional utility. However, there is no technical reason the utility should not own the service. Any barriers are defined by the company and the regulator.

A key question for the utility to ask is how far it wishes to venture into an unreg-ulated marketplace to deliver these services, and how much will it rely on other members of the ecosystem. On one hand, the financial benefits to the utility could be attractive. One recent study found that a large-scale utility-owned residential rooftop solar programme could generate a boost of 2% to 5% to shareholder earn-ings over a twenty-year period, compared to a 2% earnings loss when an equal amount of rooftop solar is owned by other parties (Barbose & Satchwell, 2020). Such a programme could also benefit all ratepayers by increasing system-wide effi-ciency of the electrical system, since sites for rooftop solar owned by the utility can be selected based on inter-connection costs and their value in deferring network up-grades (Barbose & Satchwell, 2020). Societal benefits could also accrue from utility ownership of rooftop solar, as issues of social equity and the inequitable shifting of fixed costs to lower income customers (as discussed in Chapter 10) could be ad-dressed by moving the service from a competitive marketplace to one controlled by the regulated utility (Barbose & Satchwell, 2020).

In some jurisdictions, the business of deregulated markets is isolated from the regulated utility through the establishment of a sister company, dedicated to operat-ing in deregulated markets. However, regulated utilities tend to be cautious at the prospect of becoming directly involved in markets that are not a natural monopoly, by extending the boundaries of their business beyond the customer meter. Most have not substantively altered their business model to address the anticipated growth of distributed resource. This caution exhibited by most regulated companies was re-flected by a utility executive in this most frank and candid comment:

---

3 See the glossary for definition of "demand response."

I always remind people: "You know what? Having a monopoly position is a very good business. Don't ever forget that." Competitive business can be very harsh.

Utility executive, author's research session B2

Summary

It is a fact that the Model T was only offered in black for many years, as Henry Ford focused on improving production efficiency (Alizon et al., 2009). However, after several years the Model T was eventually made available to customers in green, grey, and red, and by the end of its production run in 1927, the Model T was offered in many different colours (Alizon et al., 2009). If fact, by the end of its life, after nineteen years and fifteen million vehicles, the basic Model T had become a platform for mass customization, offering eleven different models in different colours, including a Touring Sedan, Coupe, and Runabout (Alizon et al., 2009). The Model T even formed the platform for a Ford agricultural tractor. The product that started out as ruthlessly standardized for production efficiency, evolved into a platform supporting mass customization (Alizon et al., 2009).

The value offering presented by regulated utilities for the delivery of electricity may not have the potential for mass customization that Henry Ford have found with his Model T. However, just like Henry's Model T, utility customers are increasingly being given the opportunity to customize the basic utility service with add-ons that are available on their side of the meter. The regulated utility's basic electrical service can now be integrated with services from distributed resources that offer clean energy, autonomy, cost savings (in some cases), and as electric vehicles continue to grow in market penetration, integration with their chosen method of transportation. The ecosystem is evolving (see Table 11.1), and utilities must determine their role within it.

Table 11.1: Key Issues Reshaping Value Offerings of Regulated Utilities.

Issue	Description
Grid Services	DER customers who receive services from the grid, including backup, interconnectivity, and frequency and voltage control. The grid provides these services much more economically than customers would be able to self-provision.
Backup	For DER customers who are considering migrating totally off-grid, the cost of providing backup through a generator and batteries is high relative to the cost of staying attached to the grid.
Voltage control	Most DER systems rely on grid power for voltage and frequency reference, and will not function if disconnected from the grid, or in the event of a grid outage.

Table 11.1 (continued)

Issue	Description
Interconnectivity	The grid interconnects DER producers to potential customers, providing an outlet to sell surplus electricity.
Clean Energy	A continuing connection to the grid allows risk averse customers to install distributed renewable resources, who would not otherwise do so due to the risk of service outages when wholly reliant on distributed resources. When used to support the rollout of distributed resources, the grid is also a key tool of public policy, especially as it relates to decarbonization of the economy.
Positioning in the Ecosystem	An ecosystem of firms supply customers with distributed energy services, from EV batteries and charging, to rooftop solar. Regulated utilities deliver electricity to the customer meter, but generally avoid competing in markets beyond the meter. These utilities must determine, in the growing ecosystem of unregulated suppliers to an increasingly distributed grid, in what areas of that ecosystem to compete. What elements of the ecosystem should it join as a competitive participant, and which should be left to partners and competitors?

12 The New Business Model – Centralized and Decentralized Coexistence

So, it is really a fairly new concept to us. And none of us envisioned that the customer would give us something back. And now we are looking at distributed generation, where the customer is giving us something back. So, I think a decade from now from where we are today, might be something that we haven't even fully envisioned yet.

Utility executive, author's research session F3

There is a recognition that the business model that has dominated the North American distribution utility since the 1930s is changing. Books have been written, industry forums have published studies, and there is an emerging body of scholarly work examining the subject. That a business model that has been so stable for so long might eventually come under pressure to change, should not be a total surprise. To use a quotation from Joseph Schumpeter, "any system designed to be efficient at a point in time will not be efficient over a point in time" (Garud & Karnoe, 2001, p. 6). However, as so often with changes driven by technology, the general direction of the change in the dominant business model might be estimated, but the ending location certainly has not been determined. Even more difficult to predict than direction is the timing of its arrival.

This chapter will:
- recap the current key influences impacting the utility business model
- recap the current state of the business model in transition
- describe the range of options for the future business model elements
- discuss the shape of future generic business models
- discuss a few things that we know, based on experience

Key Influences on the Business Model

The broad trends influencing North America's electric distribution networks were described in Chapter 6: decentralization, decarbonization, digitization and declining load growth. To avoid repetition, complete descriptions of these trends will not be reiterated here, but summarized:

Decentralization

Significant cost reductions over the past decade of rooftop solar (Barbose & Satchwell, 2020) have enabled growing implementation of energy resources on what has traditionally been the customer side of the meter and outside of the traditional control or ownership of the utility. In addition, batteries are increasingly present on the

https://doi.org/10.1515/9783110714036-015

customer side of the meter and are expected to grow substantially in presence as their prices continue to fall, and as electric vehicles increase their market penetration. Increased investment in centralized renewable generation will not significantly affect the business model of traditional utilities since the capital expenditure can be accommodated within the existing regulatory structure. However, increased investment in decentralized energy resources will not easily fit the regulatory structures that have been in place for decades and may disrupt the existing utility business model.

Decarbonization

In the portfolio of options that policy makers have available to them to move to a low carbon economy, the electrical grid is a key resource. The grid, which is still a major contributor to GHGs in most developed economies, will increasingly shift to low carbon generation, which in many jurisdictions will include distributed energy resources such as solar or wind generation combined with decentralized storage. In addition, the grid is an important resource to enable decarbonization of other sectors of the economy, such as transportation. To meet decarbonization goals, policy makers in many jurisdictions will require that utilities encourage the implementation of these distributed resources and build the infrastructure to support widespread charging of electric vehicles.

Digitization

The addition of digital communication to the electricity distribution network allows enhanced control and automation of the network. This enhanced communication and control is essential to bring larger amounts of distributed energy into the electrical system and to control the resultant two-way flow of electricity. At the same time, it enables increasing communication with the customer, support for new services to the customer such as EV charging, and will enable new revenue capture schemes that more closely match prices with marginal cost.

Declining Load Growth

The consumption of electricity in most industrialized nations on a per capita basis has declined in the last decade. For a high fixed cost business like a regulated utility, declining load can have particularly negative effects on the firm's margins. Declining load under a cost of service regulatory model will lead to increased rates, followed by a tendency for ratepayers to further reduce consumption, a hypothetical condition referred to as a "utility death spiral" (Graffy & Kihm, 2014, p. 2). Combined with competition from non-grid supplied electricity, the financial pressures arising from the stagnant growth of grid-supplied electricity will put increasing pressures on the profitability of the utility sector and will motivate firms to seek new business models.

The Business Model in Transition

> What is your new model you are building? Are you a wires company? Are we just allowing customers to choose where they get generation? Are we also a backup provider of last resort, which we always have been? How much procurement are we doing in the wholesale market? What is our relationship with the wholesale market, and how does that function?
>
> Utility executive, author's research session G2

With the increasing decarbonization of the grid and the economy, it appears that the utility business model is in transition, with a great deal of uncertainty about its future shape. The utility business model in transition, which was discussed in detail in chapters 8 to 11, is summarized below and in Table 12.1, with a range of potential outcomes discussed in the following section.

The Value Chain – The rapidly declining cost of distributed energy resources and the growing demand for GHG-free generation, is leading to the displacement of grid-supplied electricity in many jurisdictions. Increasingly, distributed resources that create a two-way flow of electricity are entering a grid built for centralized production and one-way flow of electricity. As illustrated in Figure 12.1, the distribution network that includes significant connections with DER becomes a "mesh," with potentially hundreds of thousands of network points and much more complexity to manage than simple one-way flow. As DER penetration grows, significant investment may be required in the grid.

The Value Offering – For decades, the customer meter has provided a natural demarcation of the utility's natural monopoly, and a boundary for the provision of services to the customer. With increasing distributed generation and batteries on the grid, the utility's boundaries of service and asset ownership are no longer clearly defined. The boundaries of resource management and ownership are being debated by industry and regulators with a wide range of possible outcomes, and the bundle of services to be offered by the utility are being redefined in many jurisdictions.

Customer Identification – The regulatory compact that was established in the 1920s and 1930s provided distribution utilities with monopoly status in an exclusive service territory, and an obligation to serve all customers within that territory. However, with customers becoming prosumers, relationships and obligations are less clear. Utilities are redefining who the customer is, and the services that the utility should be offered to that customer.

Fourth Dominant Business Model
Coexistence of Centralized and Decentralized Production

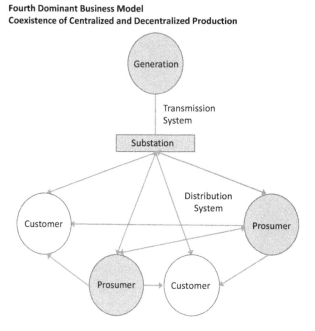

Figure 12.1: Current Business Model – In Transition.
Source: Author.

Value Capture – The growth in DER in the value chain is placing considerable stress on traditional value capture methods. For decades, the dominant business model has operated on the "cost of service" model, wherein a utility would be compensated for the investment it has made in the electric distribution system. However, since the utility's return under this model is based on its invested capital, this cost of service model does not work well when the utility does not own and manage the system assets (as is the case with many distributed assets). Furthermore, traditional bundled rates do not reflect the incremental costs of supporting consumers that move to prosumer status, resulting in a transfer of increased fixed costs to remaining customers, and an inequitable distribution of costs between customers.

The Range of Options for Business Model Elements

We have plenty of internal discussion and debate on whether or not the utility should dial all the way back to just being a core wires-operator. Or just an asset owner. We talk about whether we need to look at things like performance-based rate making. And whether or not there is going to be a bunch of energy services. When we come back and look at it, I think we believe that there will continue to be an essential function for a utility in this future. And that is particularly around the grid.   Utility executive, author's research session G1

Table 12.1: Summary of Fourth Dominant Utility Business Model (in transition).

Business Model Element	Description
Key Characteristics	Moving to a hybrid of centralized and decentralized generation. Borders between regulated and deregulated assets and services not yet determined.
Customer Identification	With customers becoming prosumers, relationship and obligations are less clear. Utilities grappling with identification of the customer, and obligations toward the customer.
Value Offering	The boundaries of the natural monopoly are no longer clearly defined due to technological innovations and declining costs of competing technologies. Since the boundaries are no longer clear, the bundle of services offered by the utility are also unclear.
Value Chain	Changes in technology and in the cost of technology are altering the distribution utility value chain. Decentralized and intermittent generation creates a more complex grid. However, digitization allows opportunities for greater control.
Value Capture	The cost of service model does not work well when the utility does not own and manage the system assets. Traditional bundled rates do not reflect the fixed costs of supporting consumers that move to prosumer status.

The dominant utility business model is in transition. Numerous organizations have proposed visions of the future shape of the business model,[1] with a broad range of potential outcomes. The purpose of this section is to describe the range of likely outcomes for the key elements of the next iteration of the business model.

---

1 Numerous organizations have proposed visions of the future of the utility business model. These include

- New York's Reforming the Energy Vision (NYPSC, 2014)
- California Public Utility Commission (Ralff-Douglas & Zafar, 2015)
- UK's RIIO Model [Revenue = Incentives + Innovation + Outputs Model] (Mandel, 2014)
- Lawrence Berkeley National Labs Models (Corneli, Kihm, & Schwartz, 2015)
- Rocky Mountain Institute's Electricity Innovation Lab Models (Cross-Call, Goldenberg, Guccione, Gold, & O'Boyle, 2018).
- MIT's Utility of the Future (Perez-Arriaga, Jenkins, & Batlle, 2017)
- Transactive Energy (multiple projects) (Melton, 2015)

This list is not exhaustive. The Energy Institute at the University of Texas at Austin has published a white paper (Abel, Agbim, et al., 2017) with an excellent summary of many of these proposed models.

The Future of the Value Chain and Value Offering

The value offerings of a future utility are closely intertwined with decisions about the utility's participation in the value chain. Hence, these two points are discussed together.

Except for the delivery of demand side management programs to customers, which was mandated by regulators starting in the 1980s and 1990s (Satchwell & Cappers, 2018), regulated utilities have historically owned and managed distribution assets only up to the customer meter, but no further. (This was discussed in Chapter 8.) Today, however, with the declining cost of non-grid supplied electricity, the range of electrical services available to customers from other suppliers is growing. (This was discussed in Chapter 11.) Despite the growth in non-grid resources, some industry commentators propose business models that would continue to restrict the utility to its traditional regulated territory, but with a role to integrate these decentralized resources (Corneli, Kihm, & Schwartz, 2015). Other commentators, on the other hand, propose future business models with the utility fully engaged in these new electrical services, with expansion of the utility's participation in the value chain to include operation and/or ownership of assets in areas that have traditionally been out of bounds. These new business models propose utility ownership of assets such as battery storage, rooftop solar, EV charging, electrical hot water heaters, and commercial grade lighting (Cross-Call et al., 2018). In one version of a proposed business model, these assets might be regulated, while in an alternative model the services could be delivered in a competitive marketplace. One example of such a proposal would see the utility own rooftop solar on a customer's premise as regulated asset and operated as a grid asset (Barbose & Satchwell, 2020). Another example of this sort of extension of the business model is provided by Green Mountain Power, a utility in Vermont, which partners with third party service providers, such as Tesla, to finance and install customer-sited batteries (Spector, 2020b). The customer can enter a ten year lease with Green Mountain for the supply and installation of two Powerwall batteries, supplied by Elon Musk's Tesla, with the customer retaining use of the battery at the end of ten years until the end of its useful life. Alternatively, the customer may install its own approved battery under a "bring your own device" option, in which the utility pays an upfront sum to the customer for purchasing a battery, but the customer makes it available to the utility as a grid-connected device. The utility, and all ratepayers, benefit by using the batteries as grid-connected assets in periods of peak demand. The customer is compensated for the battery with a storage tariff, while gaining battery backup power in times of power outage (Spector, 2020b).

The Future of Value Capture

Many discussions about new business models recognize that in an environment with growing distributed resources, traditional cost of service regulation no longer aligns the interests of the utility and ratepayers (discussed in Chapter 10). An alternative proposed in many new business models is performance based ratemaking (PBR), a regulatory framework that combines multi-year plans with specific performance targets for the utility (Cross-Call et al., 2018). The benefits of attaining those targets are often shared between the utility and the ratepayer to ensure an alignment of goals between the parties.

The use of performance goals within PBR can allow the regulator to be much more prescriptive in targeting social and environmental outcomes in addition to traditional objectives of cost and quality (Cross-Call et al., 2018). For example, the UK's RIIO system, which is probably the best known implementation of performance based ratemaking, provides utilities with traditional incentives for cost reduction, reliability, and safety (Cross-Call et al., 2018). However, the PBR structure also provides utilities with financial incentives for meeting social and environmental benchmarks, such as the provision of services to low-income customers, or the integration of low-carbon energy into the system (Mandel, 2014). Regulators using traditional cost of service regulation could not so effectively integrate social and environmental objectives directly with financial incentives. Another objective of some PBR models is the elimination of the natural incentive for a utility under cost of service regulation to incur capital costs rather than operating expenses. For example, the UK's RIIO model seeks to make the utility indifferent between these two types of expenditure by placing them both in a single regulatory class, on which the utility earns a rate of return (Spiegel-Feld & Mandel, 2015).

Proposals for updated value capture mechanisms also recognize the mismatch between utility's high fixed cost structure and the relatively small amount of fixed charge levied upon customers. The over-reliance on volumetric rates to recover fixed costs is increasingly untenable as distributed resources increase their presence on the grid. Most current proposals for value capture would see an increased recovery of fixed costs through mechanisms other than volumetric rates, such as network access fees or minimum bills (Abel et al., 2017). Proposals for value capture mechanisms also attempt to match variable rates more closely with underlying cost structure. This might include the extension of demand charges to a broader range of customers, or rates that better reflect the cost of meeting peak demand, such as time of use rates (Abel et al., 2017). It would also include altering net metering schemes to more closely reflect the true marginal benefits of the electricity provided from the distributed resource (Abel et al., 2017).

The Future of Customer Identification

The regulatory framework that exists today already has a wide variety of models for the relationship between the utility and the end consumer of electricity (discussed in Chapter 9). In a fully regulated market, much of that relationship, which would include customer service and billing, is maintained by the utility. In a deregulated market, however, most of the interface with the customer is occupied by a deregulated electricity retailer with, for example, the utility only interfacing with the customer for new service connections and trouble calls. In fact, if the utility has contracted with third parties for the provision of field and engineering services, the customer may never interface with utility staff. This broad range of options is also available in alternative business models. In fact, alternative business models could show an even deeper connection with the customer than in today's traditional regulated utility, if the utility becomes more active in managing the customers' energy experience, from in-home technologies like thermostats and management of EV charging routines. Whatever model is selected for the utility's customer identification framework, it needs to be consistent with the other elements of the business model.

Future Utility Business Models

Despite the growing attractiveness of non-grid supplied electricity, most analysis assumes that there will continue to be a benefit to customers to remain connected to the grid (as discussed in Chapter 11). Furthermore, there is a continuing value to society for utilities to encourage the integration of distributed resources into the grid, where it makes economic sense. Given these factors, and the range of options available for configuration of the business model, industry groups have made efforts to forecast the shape of the future dominant business model (Corneli, Kihm, & Schwartz, 2015). Often, these configurations fall into three generic categories:

The Energy Services Utility

The "energy services utility" would compete with other DER providers, providing grid-connected distributed assets, and support services (Corneli et al., 2015). In this case, utilities could own and operate distributed energy resources including, for example, rooftop solar, residential batteries, or EV charging infrastructure. The value capture for these services could be regulated, or the utility could operate in a competitive marketplace. The determination of whether a regulated utility should operate in an unregulated market would be a matter of public policy, reflecting government's view on the fairness of competition by regulated entities in a competitive marketplace.

The Integrating Utility

The "integrating utility" would control and coordinate, but not own, customer-sited distributed energy resources (Corneli et al., 2015) such as rooftop solar, residential batteries, or EV batteries. Utilities that have invested in systems and controls to allow these customer-owned resources to be managed as grid-resources, can manage the integrated grid as a more efficient system, managing the balancing of supply and demand more effectively. This more efficient system is beneficial to all rate payers, and the owners of the distributed resources would be compensated for the value that the distributed assets bring to the grid.

The Platform Model

Unlike a traditional "pipeline" business model, which creates and sells a product or service to a customer, the purpose of a multisided "platform" business model is to connect different sides of a market (Zhao, Von Delft, Morgan-Thomas, & Buck, 2020). These platform businesses do not take ownership of the product being sold but use other resources to bring the product to customers (Zhao et al., 2020). Traditional examples include newspapers that use classified advertisements to bring buyers and sellers of a product together, while collecting revenue from both parties. More current examples include companies that have leveraged digital technology to bring customers and suppliers together, such as Airbnb and Uber.

Some forecasters of utility business models have proposed that utilities could also utilize a platform business model for the delivery of electricity (Cross-Call et al., 2018). Under this model, the grid would be used as a platform to deliver electricity to consumers from distributed energy resources such as rooftop solar or batteries, but the utility would never take ownership of the electricity. As the operator of the platform, the utility might be compensated based on a share of the revenue, like the compensation that Uber or Airbnb would take for operating their respective platforms. Alternatively, the utility might collect a transaction fee, or a subscription charge, and might charge a fee for acting as the supplier of last resort, if the original supplier is unable to deliver the electricity as required (Abel et al., 2017). Although, this structure has not yet been tested in any significant, commercial implementations, New York State's "Reforming the Energy Vision (REV) Initiative" has developed the concept of a utility operating as a distribution system platform (DSP) (NYPSC, 2014). The New York Public Services Commission has invited utilities to propose to earn "service revenues" by hosting a platform-based marketplace for distributed energy resources. These service revenues would not totally replace traditional earnings from cost of service regulation but would be earned in parallel with cost of service regulation for traditional grid investments (Cross-Call et al., 2018).

Patterns for the Future

> Prediction is very difficult, especially about the future.   Niels Bohr (1885–1962)

It is true that prediction of the future is difficult. However, in examining the evolution of past business models, as was done in earlier chapters, there are patterns that can be useful in anticipating the nature of change in the next generation of utility business models. The following section will summarize five of these patterns that may impact changes to the utility business model:

– The transition between business models can take time, especially if firms operating under the incumbent business model have significant sunk costs.
– The "first mover advantages" enjoyed by utilities can be an impediment to change.
– Changes in utility business models may have regional differences.
– Future business models for the delivery of electricity may reflect decentralized but integrated systems. Utilities need to determine their position in the ecosystem that delivers this offering to consumers.
– Business models will change, even for entrenched monopolies.

Transition Between Business Models Can Take Time

Even if a new business model appears to be superior when introduced, transitions to the new model can take years, and sometimes decades. When Thomas Edison delivered grid-supplied electricity from his Pearl Street plant in 1882, he introduced a business model that would prove to be superior to the one being displaced, which was based on self-generation of electricity from a private plant. However, as Edison was building new central station grids across the country in the years after 1882, he continued to operate his older and more profitable business model in parallel, selling private plants (Dyer & Martin, 1910). In fact, for every central station grid Edison installed in the five years after the opening of Pearl Street, he installed ten private power plants (Bakke, 2016). In 1915, more than thirty years after the introduction of the central station grid, the most common source of electricity consumed in the United States was still the private plant, and not the electric grid (Granovetter & McGuire, 1998). The central station grid did become the country's primary source of electricity after 1915 (Granovetter & McGuire, 1998), just as Edison had envisioned in 1882, but it had taken over three decades for that business model to eventually dominate.

The transition from a business model based on DC technology to one based on AC technology also took many years. Even though AC technology had established commercial dominance by 1894 (Bakke, 2016), in 1902 there was still roughly an equal number of AC and legacy DC grids in the United States (see Chapter 5, Table 5.3). Even in the 1970s, over 300 American cities still operated a hybrid of AC systems with legacy DC systems (Cunningham, 2015).

When technology has driven the transition from one dominant business model to another in the utility sector, there has been a lengthy period of coexistence of technologies, with the sunk costs of the older technology extending the economic viability of the older business model. Although studies have found that the speed of adoption of technologies has generally accelerated over the past century (Comin, Hobijn, & Rovito, 2006), in the transition to a new utility business model that incorporates more distributed energy resources (whatever shape that may take), it seems likely that there will be a coexistence with legacy assets and legacy business models for many years.

First Mover Advantage Can Impede Business Model Change

First mover advantage describes "a firm's ability to be better off than its competitors as a result of being first to market in a new product category" (Suarez & Lanzolla, 2007, p. 122). In a competitive marketplace, when a technology is stable and market growth is smooth, first mover advantage can provide incumbents with an enviable market position (Suarez & Lanzolla, 2007). Utilities have had formidable market positions for decades thanks to being first into markets that are natural monopolies, a position reinforced with legislated market protection. This position has supported decades of financial stability, and made utilities favored investment destinations for investors seeking stable and predictable returns (Gilliland & Teufel, 2011).

However, research has also found that in some situations this same strong position can also be a disadvantage, as it may impede the incumbent firm's development of key capabilities. For example, resources and skills that are key to the growth of new market positions, such as product research and marketing, can take a long time for any firm to develop (Suarez & Lanzolla, 2007). However, at a firm with a with dominant first mover advantage in a slow growth industry, these skills are not of particular importance, and will probably remain under-developed (Suarez & Lanzolla, 2007). With their legacy of an exceptionally strong market position, many utilities are probably poorly equipped to develop the skills needed to compete in new markets. If utilities recognize their relative disadvantage in these resources and skills, they may elect to follow a conservative path by pursuing a business model focused upon traditional, regulated markets. They may elect to stick with traditional business models for as long as possible, while leaving the competitive and evolving market on the customer side of the meter to others.

Changes in Business Models May be Regional

Sometimes a business model succeeds due to regional factors. Take, for example, the early success of the business model of Edison's first central station grid. Because his DC-based technology could only distribute electricity a short distance, his central station grids were best suited for urban areas with a densely located customer base. For this reason, Edison's first station was located to serve customers in downtown Manhattan, and the economics of the business model would not have transferred easily to service the farms of rural upstate New York.

Today's utilities operate in jurisdictions with large regional differences that may support the eventual evolution of different utility business models. Consider, for example, three attributes that can be quite different between utilities: comparative electricity price, renewables penetration, and the state of regulation in that jurisdiction.

–   **Comparative Electricity Price:** Prices for electricity can vary widely between jurisdictions. In the United States and Canada, the price of electricity for residential customers varies from 5.8 US cents per kWh[2] in Montreal to 26.9 cents in New York and 29.2 cents in San Francisco (Hydro Quebec, 2020). The same type of variance is also found in Europe, with prices ranging from 0.10 Euros per kwh in Bulgaria to 0.30 Euros per kwh in Germany (Eurostat, 2021). If utility business models are changing in response to the threat of displacement of grid-supplied electricity by distributed renewables, the need for a utility to respond to this threat is far more likely in an environment like Berlin or San Francisco, where grid supplied electricity is expensive, than in a low cost environment like Sofia or Montreal, where cost advantage shields the utility from competition.

–   **Renewables Penetration:** In 2021, a new US Administration committed to a carbon free electrical grid by 2035 (Baker & Kaufman, 2020). Other developed countries have developed similar targets, albeit most with less aggressive timelines. To meet these deadlines, policy makers and regulators will require utilities over the next decade to increasingly source electricity from carbon-free generation. This will put different strains on different utilities. Take for example, a utility like Oklahoma Gas and Electric. While it does have one of the lowest prices of electricity in the United States, in 2019 approximately 94% of its generated capacity came from fossil fuels, and 6% from wind (OG&E, 2020). Hydro Quebec, as an alternative example, also has low electricity prices, but more than 99% of its electricity is the product of renewable hydroelectric generation (Hydro Quebec, 2017). (There is no ill intent to highlight either utility, except to provide a relevant example.) To meet the requirements of carbon free electrical grids that will be put in place over the next decade, regulators can be expected to place much different requirements for change on these two utilities, and the

---

2 Assumed conversion at $0.80 Cdn. per $1 US.

pressure to re-examine their business model will be much different between the two organizations.

- **State of Regulation:** Electricity regulators in different jurisdictions can drive change very differently. In California, for example, the regulator has been at the forefront of policy development supporting the deployment of distributed resources such as rooftop solar and plug-in electric vehicles, but also has some of the most expensive electricity in the country (Pyper, 2015). Texas, on the other hand, is a deregulated market with some of the lowest electricity prices in the United States (Hydro Quebec, 2020), but is still a leader in the deployment of renewable energy resources. Texas has by far the largest installed capacity of utility-scale wind generation capacity in the United States, larger than the next three largest states combined (NS Energy, 2020). However, since Texas has never implemented net metering or a feed in tariff,[3] the state has a low penetration of distributed energy resource (Pyper, 2015). Clearly, these regulatory bodies have determined more than one path to the growth of renewable energy, each with a different utility business model impact. Jurisdictions like Texas that place greater emphasis on centralized, utility-scale renewable generation will have less impact on the traditional utility business model than those that focus on distributed resources.

Just regarding these three factors, it is apparent that the pressure on a utility to alter its business model can vary significantly between different regions. And there are many other factors that create unique circumstances in different jurisdictions. For example, some utilities are investor owned while others are owned by a public body or a cooperative, some have assets measured in the billions of dollars and others just a few million, or some are fully integrated while others may be distribution utilities only.

Given the range of conditions in which utilities operate, it is quite possible that we may not immediately see the emergence of a single new dominant business model, but instead several regional variations. As studied by Klepper and Thompson (2006), some industries are actually made up of various distinct sub-markets, each with distinctive properties. The electric utility sector could very well evolve in that direction (or many directions, as the case may be), with distinct regional business models reflecting differing political and economic factors.

---

3 Net metering and feed in tariffs were discussed in Chapter 10. See the glossary for short definitions of "net metering" and "feed in tariff."

Business Models Based on an Integrated System – From Edison to Musk

One of Thomas Edison's unique strengths was his ability to think and execute in terms of integrated systems. Interviews of Edison shortly after the invention of his light bulb indicated that he saw his newly invention not as just a single device, but as part of a larger system that included electrical distribution to the customer from a central generating station (see Edison's interview in Chapter 4). While his competitors were focused on the production and sale of individual electrical components that would enable customers to generate their own electricity, Edison was already focused on the development of a complete and integrated system that would deliver lighting and electricity to the end consumer (Hughes, 1993).

Just as Thomas Edison thought in terms of systems, some of today's providers of electrical services on the customer side of the meter similarly appear to be focused on the delivery to customers of a complete system. Take for example, one competitor on the customer side of the meter that appears particularly well positioned to offer a complete system of decentralized electricity to the customer. Elon Musk's Tesla and its affiliated companies are focused on developing an integrated bundle of products that includes both energy and transportation. This would include generation of electricity from a Tesla Solar Roof, storage and backup from Tesla Powerwall batteries, connected to a Tesla EV charger, that in turn connects to a Tesla electric vehicle powered by Tesla batteries. The system could be integrated with software to respond to many external inputs, including price signals from the utility to take advantage of lower rates, forecasts of turbulent weather to top up batteries in preparation for grid outages, or changes in driver schedules to coordinate EV charging.

> If you have a great solar roof, and you have a battery pack in your house, and you have an electric car, that scales worldwide. You can solve the whole energy equation with that.
>
> Elon Musk, October 2016 (Lucchesi, 2016, p. 1)

Musk's vision does not encourage the customer to unplug from the grid, noting that, "the solution is both local power generation and utility power generation, it's not one or the other" (Lucchesi, 2016, p. 1). However, Tesla does appear to be positioning itself to manage the interface between the customer and the grid. A recent Tesla customer survey provided an indication of the company's potential role in that relationship, asking, "under what conditions would you allow Tesla to control the charging time of your car so that it is charged for your daily needs and to offer you a cheaper electricity tariff? (Lambert, 2020, p. 1)

Tesla appears to be transitioning from its origins as an automobile company that sells cars with electric motors, to an integrated energy and transportation company. The company has the potential to occupy a unique position in the marketplace, integrating a customer's on-premises generation and storage of electricity with the charging of electric vehicles, while taking control of the interface with the utility on behalf of the customer. Time will tell whether Musk and Tesla will be successful in developing a business model that offers customers a viable value proposition that includes integrated energy and transportation. The total cost of an integrated Tesla system is currently significantly more than the alternative of traditional grid-supplied electricity combined with a gasoline powered automobile. However, as discussed in Chapter 7, the costs of solar voltaics and batteries have declined rapidly over the past decade and are forecast to continue their sharp decline in the coming years.

It is possible that other firms may overtake Tesla in defining the business model for the delivery of electrical services on the other side of the customer meter. Some firms may attempt to emulate Tesla's business model with lower cost structures. Others may focus on enhancing other elements of the value offering. For example, the resilience offered by a decentralized system may become increasingly attractive to customers who have seen the experience of customers in Californian and Australian wildfires, Puerto Rican hurricanes, or Texas cold snaps. There are many potential business model outcomes. Nevertheless, what is known is that this ecosystem of suppliers of energy and transportation is becoming more complex, and utilities must determine their role in this ecosystem.

The Business Model Will Evolve – Even for Entrenched Monopolies

Earlier chapters examined the changes that occurred to utility business model elements as the industry moved from one dominant model to the next (see Table 12.2). One pattern that appears to emerge is that a substantial shift in one element of the business model will eventually require the redefinition of the other elements of the business model. When the emergence of AC technology altered the utility value chain in the 1890s, for example, it eventually forced the transformation of other business model elements, until a new utility business model was developed. Although the transformation was driven from the value chain, this new business model also reflected new value offerings, customer identification, and methods of value capture.

Take as a further example, the telecoms industry of the 1980s, which was transformed over the following few decades by a digital transformation of the value chain. Prior to its breakup into seven regional companies, AT&T held a virtual monopoly on land line service and dominated long distance telecoms service in the United States. By several measures, including assets and profitability, AT&T was the largest company in the world, employing over a million people. Furthermore, the public regarded telephone rates to be affordable and fair (Robinson, 1988) and in the early 1980s over 92% of American homes had a land line telephone (Blumberg & Luke, 2009). Few observers of the telecoms industry at the breakup of the AT&T monopoly in 1984 envisioned today's high levels of customer abandonment from what was then the core of their business, the land-line telecoms service (Kind, 2013).

Table 12.2: The Evolution of the Utility Business Model.

Business Model Description	Era	Business Model Element			
		Customer Identification	Value Offering	Value Chain	Value Capture
Pre–Utility *Localized production. Consumption.*	Pre 1882	Customer and factories large enough or individuals wealthy enough to be able to afford a dedicated power plant.	Value offering defined by the sale of electrical equipment, and subsequent maintenance or sale of replacement parts.	Companies manufacture electrical equipment for self-generation of electricity by private customers, for self-consumption.	Transactional, based on sale of equipment, installation, and replacement components.
First Dominant Utility Business Model (first introduced at Pearl Street) *The first electric grid. Localized production. Competition.*	1882 to 1890s	Customers could now be small enough to afford the installation one or two light bulbs rather than a dedicated power plant. Constraints of DC technology required customers to be limited by proximity to generating station.	**Area of the business model leading change:** The value offering represented the sale of electricity rather than the sale of equipment, a first in North America. Localized distribution of DC electricity for lighting. Driven by Edison's vision of cities electrified by a network of DC-based power-plants.	All substantial elements of the generation and distribution system, down to the very light bulb itself, was designed and manufactured by Edison.	Early attempt at volumetric pricing, with charges "by the light bulb."

			Area of the business model leading change:	
Second Dominant Utility Business Model *Centralized production. Competition.*	1890s to 1920s	Implementation of AC technology allowed customers to be defined by access to larger transmission and distribution systems, and not constrained by proximity to the generating station. Increased customer base and customer diversity allowed significant system efficiencies. For many customers, the expansion of the AC network represented their first connection to a grid. Electrification over the period up to the 1920s was accompanied by a sharp decline of costs.	With the introduction of DC technology, the value chain structure was introduced, that still exists today, of "generation-to-transmission-to-substation-to-distribution-to-customer." Steam turbine technology enabled increased plant size and efficiency, reduced unit costs.	Volumetric pricing, with charges by the kilowatt hour. Prices established by competition, or sometimes price capped by early regulation.
Third Dominant Utility Business Model *Centralized production. Regulated.*	1920s to 2010s	The regulatory compact provides the utility with monopoly status in an exclusive service territory. Utility carries an "obligation to serve" all consumers in that monopoly territory. Provision of an essential service to largely dependent consumers. In return for the provision of monopoly franchise, the utility is expected to provide quality, reliable supply of electricity at reasonable cost.	Multiple technical standards under previous model brought to new industry standards. Initially fully integrated with regulated, monopoly franchise. Deregulation in some jurisdictions have separated generation, transmission, and distribution. However, distribution continues to be treated as a monopoly.	**Area of the business model leading change:** Monopoly franchise. Regulated rate of return. Volumetric rates established by utility's cost of service. Rates represent compensation for a "bundled good," including energy, wires, and other services.

(continued)

Table 12.2 (continued)

Business Model Description	Era	Business Model Element			
		Customer Identification	Value Offering	Value Chain	Value Capture
Emergent Fourth Utility Business Model *Coexistence of centralized and decentralized production. Borders of regulated and deregulated environment not yet determined.*	Early 2010s – still emergent	Change and dislocation underway: With customers becoming prosumers, relationship and obligations are less clear. Utilities are grappling with identification of the customer, and defining obligations toward the customer.	Change and dislocation underway: The boundaries of the natural monopoly are no longer clearly defined due to technological innovations and declining costs of competing technologies. Since the boundaries are no longer clear, the bundle of services offered by the utility are also unclear.	**Area of the business model leading change:** Changes in technology and in the cost of technology are changing the distribution utility value chain. Decentralized production. Two-way flow of electricity. Digitization allowing greater control of grid.	Change and dislocation underway: Cost of service model does not work well when utility does not own and manage the system assets. Traditional bundled rates do not reflect the costs of supporting consumers that move to prosumer status.

Since that time, the industry has been overtaken by a host of new technologies, from wireless phones to the growth of the internet. Today, land lines continue to be unplugged, as fewer than 38% of Americans in 2020 lived in homes with a telephone land line (Blumberg & Luke, 2020). Although the upheaval in telecoms was driven by the transformation of a digitized value chain, the new telecoms business models that grew around the old AT&T reflected new value offerings, new customer identification, and new methods of value capture. Those telecom companies that survived from the transformation have done so only by aggressive reinvention of all elements of their business models (Kind, 2013). Utilities may soon be facing a similar situation. The transformation of the utility value chain is being enabled by new digitization capabilities and declining costs of decentralized generation and storage technologies, while being driven to change by the demands of resilience and decarbonization. Although the pressure on the business model originates from the value chain for many utilities, the effects quickly spill over to redefinition of the value offering, customer definition, and value capture, just as it did for AT&T in the 1980s. The experience of AT&T demonstrated that the business model of even a dominant monopoly can be challenged. The utility sector should take heed.

List of Abbreviations

AMI	Advanced Metering Infrastructure
CHP	Combined Heat and Power
COSR	Cost of Service Regulation
CPUC	California Public Utilities Commission
ComEd	Commonwealth Edison
DC	Direct Current
DER	Distributed Energy Resources
DLR	Dynamic Line Rating
DSM	Demand Side Management
DSP	Distribution System Platform
EV	Electric Vehicle
EMP	Electromagnetic pulse
FERC	Federal Energy Regulatory Commission
FIT	Feed-in-Tariff
GHG	Greenhouse Gas Emissions
GW	Gigawatt
GWh	Gigawatt hour
IPCC	Intergovernmental Panel on Climate Change
IPP	Independent Power Producer
IRP	Integrated Resource Plan
kW	Kilowatt
kWh	Kilowatt Hour
MW	Megawatt
MWh	Megawatt Hour
NERC	North American Electric Reliability Corporation
NREL	National Renewable Energy Laboratory
NY REV	New York State's "Reform the Energy Vision" Initiative
NWA	Non-Wires Alternatives
PBR	Performance Based Ratemaking
PG&E	Pacific Gas & Electric
PV	Photovoltaic
Ofgem	UK Office of Gas and Electricity Markets
REV	New York State's "Reform the Energy Vision" initiative
RIIO	Revenue = Incentives + Innovation + Outputs (a UK regulatory model)
ROR	Rate of Return
RPS	Renewable Portfolio Standards
TOU	Time of Use
V1G	Grid to Vehicle Charging
V2G	Vehicle to Grid Charging

https://doi.org/10.1515/9783110714036-016

Glossary

Advanced Metering According to FERC: "Advanced metering is a metering system that records customer consumption [and possibly other parameters] hourly or more frequently and that provides for daily or more frequent transmittal of measurements over a communication network to a central collection point" (Federal Energy Regulatory Commission, 2017, p. 1). Sometimes also described as a "Smart Meter" or as "Advanced Metering Infrastructure" (AMI).

Alternating Current (AC) A form of electricity in which the current alternates in direction, usually in North America at 60 times per second (i.e., 60 Hz) and in Europe at 50 times per second (i.e., 50 Hz) (ABB Group, 2018).

Black Sky Hazard A black sky hazard is an event that severely disrupts the normal functioning of critical infrastructure for long durations over a wide geographic area (EIS Council, 2021).

Capacity Factor The amount of energy a power plant actually produces, compared to the amount that it would have produced if it ran at 100% of capacity for that same period, is the capacity factor of the plant. The capacity of a plant is normally greater than the amount it actually produces in a period. Capacity factors can vary greatly between different types of generation. For example, in 2017, nuclear generation plants in the United States averaged 92% capacity factor, while wind averaged 37% and solar photovoltaic 27% (US Department of Energy, 2018).

Cost of Service Regulation In cost of service regulation, a utility's rates are set based on the total amount a utility is forecast to pay for operating expenses and capital costs, including a rate of return on the utility's equity investment in the regulated assets. Cost of service regulation may also be referred to as "rate of return" regulation (Ghadessi & Zafar, 2017).

Demand Response As defined by FERC, demand response resources include: "Changes in electric usage by demand-side resources from their normal consumption patterns in response to changes in the price of electricity over time, or to incentive payments designed to induce lower electricity use at times of high wholesale market prices or when system reliability is jeopardized" (Federal Energy Regulatory Commission, 2017). As utilities integrate more variable renewable energy into their systems (e.g., wind and solar), demand response programs may become more important to help utilities manage the electrical system.

Demand Side Management The purpose of an electric utility's "demand-side management" (DSM) programs is to alter their customers' use of electricity in ways that will change the shape of the utility's load shape. DSM includes all demand management measures, including demand response, and energy efficiency.

Dispatchable Electricity Dispatchable electricity refers to electricity that is available to a power system within the time required by a system operator. Utilities maintain certain levels of dispatchable generation capacity that is unloaded but available to respond on a timely basis to system requirements. Different resources have different dispatch response times from a cold start, with capacitors and hydroelectric facilities measured in seconds, gas turbines measured in minutes, and coal and nuclear stations measured in hours. Renewable energy sources, like solar and wind, are not treated as dispatchable by many utilities due to their intermittence.

Distributed Energy Resource Guidelines set out by the New York Independent System Operator (DNV GL Energy, 2014, p. 1) define DER technologies as:

https://doi.org/10.1515/9783110714036-017

"behind-the-meter" power generation and storage resources, typically located on an end-use customer's premises and operated for the purpose of supplying all or a portion of the customer's electric load, and may also be capable of injecting power into the transmission and/or distribution system, or into a non-utility local network in parallel with the utility grid. These DERs includes such technologies as solar PV (photo-voltaic), CHP (combined heat and power) or cogeneration systems, microgrids, wind turbines, micro turbines, back-up generators and energy storage.

These guidelines note that Demand Response programs are assumed not to be within the scope of DER.

Distribution See "Transmission and Distribution."

Distribution Substation A distribution substation is the transition point for moving electrical power from a high-voltage transmission system to a lower voltage distribution system, from where it is distributed to customers connected to the distribution system. For a distribution utility, the distribution substation is the typical boundary of its business, with transmission operators and generation utilities operating upstream of the substation.

Dynamo A dynamo is an electro-mechanical device that generates direct current (DC) electric power. It is a form of generator, although the term "generator" normally refers to an electro-mechanical device that generates alternating current (AC) electric power. A generator that produces AC electric power is also referred to as an "alternator."

Electric Utility The US Department of Energy defines a utility as a legal entity "aligned with distribution facilities for delivery of electric energy for use primarily by the public" (US Department of Energy, 2018f, p. 249). These may include investor-owned electric utilities, municipal and state utilities, federal electric utilities, and rural electric cooperatives. The definition notes that since the traditional electric utilities have started to functionally unbundle their generation, transmission, and distribution operations, "electric utility" currently has inconsistent interpretations between jurisdictions (US Department of Energy, 2018f).

Feeder Feeders are electrical cables that connect consumers of electricity to distribution substations.

Feed in tariff A tariff paid to a generator of electricity for electricity fed into the grid from a distributed resource. This tariff is often designed to incentivize investments in renewable energy resources (e.g., solar, wind). Net metering, on the other hand, offsets a customer's electricity consumed against electricity that the customer has generated and fed back into the grid. Accordingly, under net metering, the producer of electricity is compensated at the same tariff rate as the electricity consumed.

Grid-interactive inverter A grid-interactive inverter (see definition of inverter) performs the same functions as an inverter, but with several additional features. When grid supplied electricity is interrupted, a grid-interactive inverter automatically activates to supply DC power from distributed energy sources, such as solar panels or batteries, into useable AC power for household loads.

Grid parity Grid parity can have several meanings, but most often refers to the point at which an alternative energy resource is available to a consumer at a levelized cost that is equal to or less than the cost of grid-supplied electricity.

Independent power producer (IPP) A generator that is independent from the utility, may sell electricity to the utility, but is not itself part of the electric utility.

Integrated Resource Plan (IRP) An Integrated Resource Plan lays out the utility's plan to meet future growth in demand for electricity through energy conservation and generation.

Inverter An inverter is an electrical device that converts DC electrical current (electricity generated by distributed energy resources such as wind turbines or photovoltaic solar panels is typically DC current) into AC electrical current (electricity distributed within households or on the electrical grid is typically AC current). Also see "Smart Inverter" and "Grid-interactive inverter."

Kilowatt hours A measure equivalent to the consumption of 1,000 watts of electricity for 1 hour. Abbreviated as kWh, it forms the typical unit of measurement used in volumetric charging for electricity consumption. See definition of watt.

Line loss Line loss is the electric energy lost in the transmission of electricity through an electrical grid before it is delivered to the customer. Much of the energy loss is due to heating of the cables carrying the electrical current. A line loss is generally an economic loss since the energy is expended without economic benefit to the customer.

Load The load of an electrical power system is the total amount of electricity consumed by all users on that system. The term also refers to the amount of electrical power consumed by a particular device (ABB Group, 2018).

Microgrid A microgrid is a localized grid that can disconnect from the traditional grid to operate autonomously. Since microgrids are able to continue to operate while the main grid is down, microgrids can strengthen grid resilience and mitigate grid disturbances (US Department of Energy, 2011).

Net Metering See definition of "feed-in-tariff (FIT)."

Performance Based Ratemaking Performance-based regulation (PBR) is an alternative regulatory framework designed to align the interests of utilities with those of its ratepayers and the public. PBR provides financial incentives to utilities for achieving performance targets established in a multi-year rate plan.

Photovoltaic energy Electricity generated from sunlight through solid-state semiconductor devices that have no moving parts (US Department of Energy, 2018f).

Rate Base The undepreciated cost of the investment that a public utility has made in regulated assets used to deliver service to the customer. It is on this undepreciated cost that a public utility is permitted to earn a specified rate of return, in accordance with rules set by a regulatory agency. In general, the rate base consists of the value of property as used by the utility in providing service (Ghadessi & Zafar, 2017).

Rate Case The rate case process is used by the utility and the regulator to establish the electricity prices that can be charged for different customer classes in a period. These prices are determined through forecasts of the utility's revenue requirements and forecast consumption of electricity. The interval of rate case can vary between single years to multiple years (Bird, Hurlbut, Donohoo, Cory, & Kreycik, 2009).

Rate of Return Regulation A form of regulation that allows a firm to set its revenues to allow the firm to recover the cost of delivering a regulated service, plus a rate of return on the equity that the firm has invested in assets used to deliver that regulated service. Also described as "Cost of Service Regulation."

Smart Grid A smart grid is an electrical grid that utilizes information technology to manage electricity networks to improve reliability, cost effectiveness and energy efficiency (ABB Group, 2018).

Smart Meter See "Advanced Metering."

Smart Inverter A smart inverter (see definition of inverter) is an inverter attached to a distributed energy source (such as rooftop solar) which is controllable by the utility, or other third party, to remotely mitigate voltage change (such as those caused by fluctuations in solar power) or other factors.

Transformers Transformers are devices that transfer electricity from one circuit to another, raising or lowering the voltage of the electrical current while leaving frequency unchanged. Changing the voltage of a current as it is transported is performed to increase the efficiency of power systems, and for reasons of safety. There are no moving parts in a transformer, so the economic life of a transformer deployed in a distribution system is often measured in decades.

Transmission and Distribution "Transmission" is the movement of power at high voltage, usually over long distances, generally from the point of generation to a substation. Power is transmitted at high voltage for greater efficiency and to enable transmission over longer distances. Very large customers, such as large industrial plants, may receive their electric power directly as transmission customers, although most customers will receive their electric power through the distribution system. "Distribution" is the transport of electricity at medium voltage over shorter distances to end consumers. The demarcation line between distribution and transmission is generally the distribution substation (ABB Group 2018).

Utility: See definition of "Electric Utility."

Watt, kilowatt, megawatt, and gigawatt A watt is a unit of power, while kilowatt is 10^3 watts and a megawatt is 10^6 watts, and a gigawatt is 10^9 watts. Watt-hours are measurements of energy, describing the amount of electricity consumed in one hour. Typically, volumetric rates for electricity customers are charged in kilowatt hours. An electric appliance rated at 1,000 watts, such as a typical microwave, running for one hour, will consume one kilowatt hour of electricity. A typical Canadian household uses about 11,000-kilowatt hours kWh of electricity per year (Statistics Canada, 2007). A typical American household uses about 10,766 kWh hours of electricity per year (US Energy Information Administration, 2018b).

References

Introduction

Berry, J. (1988, January 19). Public Service of N. H. Files for Chapter 11. *The Washington Post*. Retrieved from: https://www.washingtonpost.com/archive/business/1988/01/29/public-service-of-nh-files-for-chapter-11/891cd39e-c273-4458-9a76-2697a87c27b3/ (Accessed 2021 January 7)

Clean Energy Wire (2018, July 10). "Big four" German utilities drop out of top 100 company list. *Clean Energy Wire*. Retrieved from: https://www.cleanenergywire.org/news/renewables-overtake-coal-sino-german-cooperation/big-four-german-utilities-drop-out-top-100-company-list (Accessed 2021 June 5)

Constable, G., & Somerville, B. (Eds.). (2003). *A Century of Innovation: Twenty Engineering Achievements That Transformed Our Lives*. Joseph Henry Press.

Howe, C. (2016 July 22). Infrastructure: Power to the people. *Science*, 353(6297).

Introduction to Part 1

Bakke, G. (2016). *The Grid: The Fraying Wires Between Americans and our Energy Future*. Bloomsbury Publishing.

Henao, L., & Byrne, P. (2019 June 16), Blackout in South America raises questions about power grid. *The Associated Press*.

Lovins, A. B., & Lovins, L. H. (1982). *Brittle Power: Energy Strategy for National Security*. Brickhouse Publishing Company, Andover.

Chapter 1

Amelang S., Appunn K., & Wettengel J. (2020 October 5). Europe's 55% Emissions Cut by 2030: Proposed Target Means Even Faster Coal Exit. energypost.eu Retrieved from: https://energypost.eu/europes-55-emissions-cut-by-2030-proposed-target-means-even-faster-coal-exit/ (Accessed 2021 March 29)

BloombergNEF (2020a December 16). Battery Pack Prices Cited Below $100/kWh for the First Time in 2020, While Market Average Sits at $137/kWh. *BloombergNEF*. Retrieved from: https://about.bnef.com/blog/battery-pack-prices-cited-below-100-kwh-for-the-first-time-in-2020-while-market-average-sits-at-137-kwh/ (Accessed 2021 January 24)

Canizales, A. (2021 March 1). Lt. Gov. Dan Patrick calls for resignations at Public Utility Commission. *Texas Tribune*. Retrieved from: https://www.texastribune.org/2021/03/01/dan-patrick-texas-ercot-resign/ (Accessed 2021 March 2)

Cinnamon, B. (2019 December 2).The Myth of Whole-Home Battery Backup. *GreenTech Media*. Retrieved from: https://www.greentechmedia.com/articles/read/the-myth-of-whole-home-battery-backup (Accessed 2020 August 5)

Coto, D. (2020). Report: FEMA fumbled in Puerto Rico after storms Irma, Maria. *Associated Press*. Retrieved from: https://apnews.com/article/puerto-rico-hurricane-irma-storms-latin-america-hurricanes-8bfd2865519e79a2109bab7edb625436 (Accessed 2021 March 23)

https://doi.org/10.1515/9783110714036-018

Deign, J. (2020 March 31). Germany's Maxed-Out Grid Is Causing Trouble Across Europe. *GreenTech Media*. Retrieved from: https://www.greentechmedia.com/articles/read/germanys-stressed-grid-is-causing-trouble-across-europe (Accessed 2021 March 29)

EIS Council. (2021 January 28). Black Sky Hazards. The Electric Infrastructure Security (EIS) Council. Retrieved from: https://www.eiscouncil.org/BlackSky.aspx (Accessed 2021 January 28)

Feldman, D., Ramasamy, V., Fu, R., Ramdas, A., Desai, J., & Margolis, R. (2021). *US Solar Photovoltaic System and Energy Storage Cost Benchmark: Q1 2020* (No. NREL/TP-6A20-77324). National Renewable Energy Lab. (NREL), Golden, CO.

Gerdes, J. (2019, November 8). Will Your EV Keep the Lights on When the Grid Goes Down? *GreenTech Media*. Retrieved from: https://www.greentechmedia.com/articles/read/will-your-ev-keep-the-lights-on-when-the-grid-goes-down (Accessed 2021 March 28)

Gramlich, R. (2021, February 18). No state is an island – Transmission keeps the lights on. Utility Dive. Retrieved from: https://www.utilitydive.com/news/no-state-is-an-island-transmission-keeps-the-lights-on/595291/ (Accessed 2021 March 29)

Hay, A. (2019, November 22). Explainer: California faces decade of 'unique' wildfire blackouts. Reuters. Retrieved from: https://www.reuters.com/article/us-california-wildfire-pg-e-explainer-idUSKBN1XW1AF (Accessed 2021 March 2)

IPCC (Intergovernmental Panel on Climate Change). (2018a). IPCC Press Release: Summary for Policymakers of IPCC Special Report on Global Warming of 1.5°ºC approved by governments. Retrieved from: https://www.ipcc.ch/2018/10/08/summary-for-policymakers-of-ipcc-special-report-on-global-warming-of-1-5c-approved-by-governments/ (Accessed 2021 March 2)

IPCC (Intergovernmental Panel on Climate Change). (2018b). *Global Warming of 1.5°C, Summary for Policy Makers*. Retrieved from: http://ipcc.ch/report/sr15/ (Accessed 2021 March 2)

Jenkins, J. (2021 March 18). Testimony of Dr. Jesse D. Jenkins, Committee on Science, Space and Technology, United States House of Representatives. Retrieved from: https://www.dropbox.com/s/y1zw97dbq264aj3/Jenkins_House_Science_and_Technology_Committee_testimony_031821.pdf?dl=0 (Accessed 2021 April 6)

Kwasinski, A., Andrade, F., Castro-Sitiriche, M.J., & O'Neill-Carrillo, E. (2019). Hurricane Maria effects on Puerto Rico electric power infrastructure. *IEEE Power and Energy Technology Systems Journal*, 6(1), 85–94.

Larson, E., Greig, C., Jenkins, J., Mayfield, E., Pascale, A., Zhang, C., & Drossman, J. (2020). Net-Zero America: Potential Pathways, Infrastructure and Impacts. Informe provisional. Princeton, NJ: Princeton University. Retrieved from: https://environmenthalfcentury.princeton.edu/sites/g/files/toruqf331/files/202012/Princeton_NZA_Interim_Report_15_Dec_2020_FINAL.pdf (Accessed 2021 April 4)

MacDonald, A. E., Clack, C. T., Alexander, A., Dunbar, A., Wilczak, J., & Xie, Y. (2016). Future cost-competitive electricity systems and their impact on US CO2 emissions. *Nature Climate Change*, 6(5), 526–531.

Merchant, E. F. (2018, October 8). IPCC: Renewables to Supply 70%-85% of Electricity by 2050 to Avoid Worst Impacts of Climate Change. *GreenTech Media*.

Milman, O., Chang, A., & Kamal, R. (2021 March 15). The race to zero: Can America reach net-zero emissions by 2050? *The Guardian Newspaper*. Retrieved from: https://www.theguardian.com/us-news/2021/mar/15/race-to-zero-america-emissions-climate-crisis# (Accessed 2021 April 7)

Morton, A. (2020, November 5). Victoria plans 300MW Tesla battery to help stabilise grid as renewables increase. *The Guardian Newspaper*. Retrieved from: https://www.theguardian.com/australia-news/2020/nov/05/victoria-plans-300mw-tesla-battery-to-help-stabilise-grid-as-renewables-increase (Accessed 2021 April 4)

Mulcahy, S. (2021, February 19). Many Texans have died because of the winter storm. *Texas Tribune*. Retrieved from: https://www.texastribune.org/2021/02/19/texas-power-outage-winter-storm-deaths/ (Accessed 2021 March 2)

Narang, D., Ingram, M., Li, X., Stout, S., Hotchkiss, E., Bhat, A., . . ., & Latif, A. (2021). Considerations for Distributed Energy Resource Integration in Puerto Rico: DOE Multi-Lab Grid Modeling Support for Puerto Rico; Analytical Support for Interconnection and IEEE Std 1547-2018 National Renewable Energy Laboratory (Task 3.0). (No. NREL/TP-5D00-77127). National Renewable Energy Lab (NREL), Golden, CO. Retrieved from: https://doi.org/10.2172/1769814 (Accessed 2021 April 5)

NASEM (National Academies of Sciences, Engineering, and Medicine). (2017). *Enhancing the Resilience of the Nation's Electricity System*. National Academies Press.

Prakash, K., Lallu, A., Islam, F. R., & Mamun, K. A. (2016, December). *Review of Power System Distribution Network Architecture*. In 2016 3rd Asia-Pacific World Congress on Computer Science and Engineering (APWC on CSE) (pp. 124–130). IEEE.

Presidential Policy Directive (2013 February 12). Presidential Policy Directive, Critical Infrastructure Security and Resilience. The White House. Retrieved from: http://www.whitehouse.gov/the-press-office/2013/02/12/presidential-policy-directive-critical-infrastructure-security-and-resil (Accessed 2021 March 25)

Rickerson, W., Gillis, J., & Bulkeley, M. (2019). *The Value of Resilience for Distributed Energy Resources: An Overview of Current Analytical Practices*. National Association of Regulatory Utility Commissioners.

Roberson, D., Kim, H. C., Chen, B., Page, C., Nuqui, R., Valdes, A., . . ., & Johnson, B. K. (2019). Improving grid resilience using high-voltage DC: strengthening the security of power system stability. *IEEE Power and Energy Magazine*, 17(3), 38–47.

Rust, M., & Kim, K. (2021 February 19). Why Cold Weather Cut the Power in Texas. *The Wall Street Journal*. Retrieved from: https://www.wsj.com/articles/why-cold-weather-cut-the-power-in-texas-11613765319#: (Accessed 2021 June 13)

Sanger, D., & Mazzetti, M. (2016 February 16). US Had Cyberattack Plan if Iran Nuclear Dispute Led to Conflict. *The New York Times*. Retrieved from: https://www.nytimes.com/2016/02/17/world/middleeast/us-had-cyberattack-planned-if-iran-nuclear-negotiations-failed.html (Accessed 2021 June 13)

Saqib, S. (2019). HVDC: A Building Block for a Resilient, Flexible and Interconnected Grid. *T&D World*. Retrieved from: https://www.tdworld.com/overhead-transmission/article/20972318/ (Accessed 2020 September 14)

Taleb, N. (2007). *The Black Swan: The Impact of the Highly Improbable*. Random House Incorporated.

Templeton, B. (2020 September 22). Tesla 'Battery Day' Promises 56% Reduction in Battery Cost and Much More. *Forbes Magazine*. Retrieved from: https://www.forbes.com/sites/bradtempleton/2020/09/22/tesla-battery-day-promises-56-reduction-in-battery-cost-and-much-more/?sh=1b3ee2176253 (Accessed 2021 June 10)

The Economist (2021 February 10). Joe Biden's Climate-Friendly Energy Revolution. *The Economist Magazine*. 438 (9233). Retrieved from: https://www.economist.com/briefing/2021/02/20/joe-bidens-climate-friendly-energy-revolution (Accessed 2021 June 10)

Tsuchida, T., & Gramlich, R. (2019). *Improving Transmission Operation with Advanced Technologies: A Review of Deployment Experience and Analysis of Incentives*. Brattle Group White Paper.

Trabish, H. (2021 March 2). Texas must increase ties to the national grid and DER to avoid another power catastrophe, analysts say. *Utility Dive*. Retrieved from: https://www.utilitydive.com/

news/texas-must-increase-ties-to-the-national-grid-and-der-to-avoid-another-powe/595845/
(Accessed 2021 March 28)

US EIA. (2020c). *Use of Energy Explained, Energy Use in Homes*. US Energy Information
Administration. Retrieved from: https://www.eia.gov/energyexplained/use-of-energy/homes.
php (Accessed 2021 January 9)

Wender, B. A., Morgan, M. G., & Holmes, K. J. (2017). Enhancing the resilience of electricity
systems. *Engineering*, 3(5), 580–582.

Chapter 2

Abdul Aziz, S., Fitzsimmons, J. R., & Douglas, E. J. (2008). Clarifying the Business Model Construct.
Proceedings of the 5th Australian Graduate School of Entrepreneurship International
Entrepreneurship Research Exchange, 795–813.

Arthur, W. B. (1989). Competing technologies, increasing returns, and lock-in by historical events.
The Economic Journal, 99(394), 116–131.

Baden-Fuller, C., & Mangematin, V. (2013). Business models: A challenging agenda. *Strategic
Organization*, 11(4), 418–427.

Baden-Fuller, C., & Morgan, M. S. (2010). Business models as models. *Long Range Planning*, 43
(2–3), 156–171.

Bohnsack, R., Pinkse, J., & Kolk, A. (2014). Business models for sustainable technologies: Exploring
business model evolution in the case of electric vehicles. *Research Policy*, 43(2), 284–300.

Cardwell, D. (2017 March 13). Solar Experiment Lets Neighbors Trade Energy Among Themselves.
The New York Times. Retrieved from: https://www.nytimes.com/2017/03/13/business/energy-
environment/brooklyn-solar-grid-energy-trading.html (Accessed 2021 January 29)

Chesbrough, H., & Rosenbloom, R. S. (2002). The Role of the Business Model in Capturing Value
from Innovation: Evidence from Xerox Corporation's Technology Spin-Off Companies. *Industrial
and Corporate Change*, 11(3), 529–555.

Christensen, C. M. (1997). The innovator's dilemma: When new technologies cause great firms to
fail. *Harvard Business School Press*.

David, P. A. (1985). Clio and the Economics of QWERTY. *The American Economic Review*, 75(2),
332–337.

Eisenmann, T., Parker, G., & Van Alstyne, M. W. (2006). Strategies for two-sided markets. *Harvard
Business Review*, 84(10), 92.

Fehrer, J. A., Woratschek, H., & Brodie, R. J. (2018). A systemic logic for platform business models.
Journal of Service Management. 29 (4), 546–568

Gassmann, O., Frankenberger, K., & Csik, M. (2014). *The Business Model Navigator: 55 Models That
Will Revolutionise Your Business*. Pearson UK.

Granovetter, M., & McGuire, P. (1998). The making of an industry: Electricity in the United States.
The Sociological Review, 46(S1), 147–173.

IBM Global CEO Study Team. (2006). *Expanding the Innovation Horizon: The Global CEO Study
2006*. IBM Global Services.

Karim, S., & Mitchell, W. (2000). Path-dependent and path-breaking change: Reconfiguring
business resources following acquisitions in the US medical sector, 1978–1995. *Strategic
Management Journal*, 21(10–11), 1061–1081.

Lewis, M. (1999). *The New Thing: A Silicon Valley Story*. WW Norton & Company.

Magretta, J. (2002). Why business models matter. *Harvard Business Review*, 80(5), 86.

Massa, L., Tucci, C. L., & Afuah, A. (2017). A critical assessment of business model research. *Academy of Management Annals*, 11(1), 73–104.

Matzler, K., Bailom, F., von den Eichen, S. F., & Kohler, T. (2013). Business model innovation: Coffee triumphs for Nespresso. *Journal of Business Strategy*.

McGrath, R. G. (2010). Business models: A discovery driven approach. *Long Range Planning*, 43 (2–3), 247–261.

Osterwalder, A., & Pigneur, Y. (2010). *Business Model Generation: A Handbook for Visionaries, Game Changers, and Challengers*. John Wiley & Sons.

Parker, G. G., Van Alstyne, M. W., & Choudary, S. P. (2016). *Platform Revolution: How Networked Markets Are Transforming the Economy and How to Make Them Work for You*. WW Norton & Company.

Pierson, P. (2000). Increasing returns, path dependence, and the study of politics. *American Political Science Review*, 94(2), 251–267.

Porac, J. F., Thomas, H., & Baden-Fuller, C. (1989). Competitive Groups as Cognitive Communities: The case of Scottish knitwear manufacturers. *Journal of Management Studies*, 26(4), 397–416.

Prahalad, C. K., & Bettis, R. A. (1986). The dominant logic: A new linkage between diversity and performance. *Strategic Management Journal*, 7(6), 485–501.

Sinclair, H. (2020 August 18). Are virtual power plants the future of solar power? Australia Broadcasting Corporation News. Retrieved from: https://www.abc.net.au/news/2020-08-19/virtual-power-plants-the-future-of-solar-power/12570544 (Accessed 2021 January 29)

Sosna, M., Trevinyo-Rodríguez, R. N., & Velamuri, S. R. (2010). Business model innovation through trial-and-error learning: The Naturhouse case. *Long Range Planning*, 43(2–3), 383–407.

Sull, D. N. (1999a). The dynamics of standing still: Firestone tire & rubber and the radial revolution. *Business History Review*, 73(3), 430–464.

Sull, D. N. (1999b). Why Good Companies Go Bad and How Great Managers Remake Them. *Harvard Business Press*.

Sydow, J., Schreyogg, G., & Koch, J. (2009). Organizational path dependence: Opening the black box. *Academy of Management Review*, (4), 689–709.

Teece, D. J. (2010). Business models, business strategy and innovation. *Long Range Planning*, 43 (2–3), 172–194.

Tripsas, M., & Gavetti, G. (2000). Capabilities, cognition, and inertia: Evidence from digital imaging. *Strategic Management Journal*, 21(10–11), 1147–1161.

Zott, C., & Amit, R. (2010). Business model design: An activity system perspective. *Long Range Planning*, 43(2–3), 216–226.

Introduction to Part 2

McCraw, T. K. (2006). Schumpeter's business cycles as business history. *Business History Review*, 80(2), 231–261.

Nelles, H. V. (2003). Hydro and after: The canadian experience with the organization, nationalization and deregulation of electrical utilities. *Annales Historiques de l'Electricite*, (1), 117–132.

Schumpeter, J. A. (1939). *Business Cycles: A Theoretical, Historical, and Statistical Analysis of the Capitalist Process*. McGraw-Hill. Retrieved from: https://archive.org/details/in.ernet.dli.2015.223551/page/n453/mode/2up (Accessed 2020 June 15)

US Department of Energy. (2018f). *US Energy Information Administration. Electric Power Monthly with Data for February 2018*. US Department of Energy, Washington DC. Retrieved from: https://www.eia.gov/electricity/monthly/ (Accessed 2019 July 10)

Wasik, J. F. (2006). *The Merchant of Power: Sam Insull, Thomas Edison, and the Creation of the Modern Metropolis*. Macmillan.

Chapter 3

Allan, S., & Pang, J. (2020, April). Lessons Learned from Hawaii. *Public Utilities Fortnightly*. Retrieved from: www.fortnightly.com/fortnightly/2020/04/ (Accessed 2020 November 15).

Allerhand, A. (2019). Early AC power: The first long-distance lines. *IEEE Power and Energy Magazine*, 17(5), 82–90.

Bakke, G. (2016). *The Grid: The Fraying Wires Between Americans and our Energy Future*. Bloomsbury Publishing.

Chandler Jr., A. D. (1993). *The Visible Hand*. Harvard University Press.

Dyer, T. C., & Martin, F. L. (1910). *Edison, His Life and Inventions*. New York: Harper & Brothers. Retrieved from: http://gutenberg.org/files/820/820-h/820-h.htm (Accessed 2020 November 14)

Electricity Council. (1987). *Electricity Supply in the United Kingdom: Chronology: From the Beginnings of the Industry to 31 December 1985*. Electricity Council.

Granovetter, M., & McGuire, P. (1998). The making of an industry: Electricity in the United States. *The Sociological Review*, 46(S1), 147–173.

Hargadon, A. B., & Douglas, Y. (2001). When innovations meet institutions: Edison and the design of the electric light. *Administrative Science Quarterly*, 46(3), 476–501.

Jenkins, R. V. (1982). Pearl street perspectives, *IEEE Power Engineering Review*, PER-2, (4), 9–10. Retrieved from: www.ieeexplore.ieee.org (Accessed 2020 September 05).

Jonnes, J. (2004). *Empires of Light: Edison, Tesla, Westinghouse, and the Race to Electrify the World*. Random House.

Muir, A., & Lopatto, J. (2004). Final Report on the August 14, 2003 Blackout in the United States and Canada: Causes and Recommendations. US–Canada Power System Outage Task Force, Canada.

Munson, R. (2005). *From Edison to Enron: The Business of Power and What It Means for The Future of Electricity*. Northeast-Midwest Institute.

National Research Council. (2010). America;s Energy Future: Technology and Transformation. National Academies Press.

Skjong, E., Rodskar, E., Molinas Cabrera, M. M., Johansen, T. A., & Cunningham, J. (2015). The marine vessel's electrical power system: From its birth to present day. *Proceedings of the IEEE*, 103(12),2410–2424.

Smil, V. (2017). *Energy: A Beginner's Guide*. Simon and Schuster.

Stross, R. E. (2008). *The Wizard of Menlo Park: How Thomas Alva Edison Invented the Modern World*. Broadway Books.

Sulzberger, C. (2010). Thomas Edison's 1882 Pearl Street Generating Station. *IEEE Global History Network*.

US–Canada Power System Outage Task Force, Abraham, S., Dhaliwal, H., Efford, R. J., Keen, L. J., McLellan, A., . . ., & Wood, P. (2004). Final Report on the August 14, 2003 Blackout in the United States and Canada: Causes and Recommendations. US–Canada Power System Outage Task Force.

Wasik, J. F. (2006). *The Merchant of Power: Sam Insull, Thomas Edison, and the Creation of the Modern Metropolis*. Macmillan.

Whitehead, A. N. (1911). *An Introduction to Mathematics*. New York: Henry Holt and Company.

Chapter 4

Bakke, G. (2016). *The Grid: The Fraying Wires Between Americans and our Energy Future*. Bloomsbury Publishing.

Coltman, J. W. (2002). The transformer [historical overview]. *IEEE Industry Applications Magazine*, 8 (1), 8–15.

Conot, R. E. (1979). A streak of luck. New York: Seaview Books.

Cunningham, J. J. (2015). STARS: Manhattan Electric Power Distribution, 1881–1901. *Proceedings of the IEEE*, 103(5), 850–858.

Dyer, T. C., & Martin, F. L. (1910). *Edison, His Life and Inventions*. New York: Harper & Brothers. Retrieved from: http://gutenberg.org/files/820/820-h/820-h.htm (Accessed 2020 November 15)

Granovetter, M., & McGuire, P. (1998). The making of an industry: Electricity in the United States. *The Sociological Review*, 46(S1), 147–173.

Hargadon, A. B., & Douglas, Y. (2001). When innovations meet institutions: Edison and the design of the electric light. *Administrative Science Quarterly*, 46(3), 476–501.

Heun, R. C., & Moss, H. R. (1987). Thomas A. Edison's adventures in concrete. *Concrete International*, 9(1), 12–14.

Hughes, T. P. (1993). *Networks of Power: Electrification in Western* Society, *1880–1930*. JHU Press.

Jenkins, R. V. (1982). Pearl street perspectives. *IEEE Power Engineering Review*, PER-2(4), 9–10, April 1982. Retrieved from: URL: http://ieeexplore.ieee.org

Jonnes, J. (2004). *Empires of Light: Edison, Tesla, Westinghouse, and the Race to Electrify the World*. Random House.

Millard, A. (1992). Thomas Edison, the battle of the systems and the persistence of direct current. *Material Culture Review*, 6(1). Retrieved from: https://journals.lib.unb.ca/index.php/MCR/article/view/17509/22460 (Accessed 2021 June 12)

Molinas, M., & Monti, A. (2017). The marine electrical revolution: Battery power at sea [about this issue]. *IEEE Electrification Magazine*, 5(3), 2–3.

Nelles, H. (2014). Light switch: Towards a history of the second enlightenment. Scientia Canadensis: *Canadian Journal of the History of Science, Technology and Medicine/Scientia Canadensis: Revue Canadienne d'Histoire des Sciences, des Techniques et de la Médecine*, 37(1–2), 11–33.

Noberni, R. (1982). Pearl street station – The first truly complete and integrated electric utility system. *IEEE Power Engineering Review*, PER-2(1), 2–3.

Roguin, A. (2004). Nikola Tesla: The man behind the magnetic field unit. *Journal of Magnetic Resonance Imaging: An Official Journal of the International Society for Magnetic Resonance in Medicine*, 19(3), 369–374.

Rutgers University. (2020a). Edison's newest marvel. Anonymous article in the *New York Sun*, New York, September 16, 1878, *Thomas A Edison Papers*. Retrieved from: http://edison.rutgers.edu/yearofinno/EL/Doc1439_NYSun_9-16-78.pdf.

Rutgers University. (2020b). *Thomas A Edison Papers*. Retrieved from: www.edison.rutgers.edu/latimer/tae1.htm (Accessed 2020 November 7)

Rutgers University. (2020c). Cement patents. *Thomas A Edison Papers*, Retrieved from: www.edison.rutgers.edu/cemepats.htm (Accessed 2020 September 7).

Rutgers University. (2020d). Anonymous Article in the *New York Herald*, New York, September 5, 1882, *Thomas A Edison Papers*. Retrieved from: www.edison.rutgers.edu/yearofinno/EL/ Doc2338_PearlStreetarticle_9-5-80.pdf (Accessed 2020 November 7)

Skrabec Jr., Q. R., & Skrabec, Q. R. (2007). *George Westinghouse: Gentle Genius*. Algora Publishing.

Stana, G., & Apse-Apsitis, P. (2015). An Insight into the Evolution of Direct Current Systems. In: *Closing Conference of the Project "Doctoral School of Energy and Geotechnology II,"* Estonia, Pärnu, 12–17 January 2015. Pärnu: Elektriajam, 2015, pp. 62–66. Retrieved from: http://egdk. ttu.ee/files/parnu2015/Parnu_2015_062-066.pdf (Accessed 2020 July 6).

Sulzberger, C. (2010). Thomas Edison's 1882 Pearl Street Generating Station. *IEEE Global History Network*.

Wasik, J. F. (2006). *The Merchant of Power: Sam Insull, Thomas Edison, and the Creation of the Modern Metropolis*. Macmillan.

Chapter 5

Allerhand, A. (2017). A contrarian history of early electric power distribution. *Proceedings of the IEEE*, 105(4), 768–778.

Allerhand, A. (2019). Early AC power: The first long-distance lines. *IEEE Power and Energy Magazine*, 17(5), 82–90.

Ardelean, M., & Minnebo, P. (2015). *HVDC Submarine Power Cables in the World*. Joint Research Center, European Commission.

Bakke, G. (2016). *The Grid: The Fraying Wires Between Americans and our Energy Future*. Bloomsbury Publishing.

Bradley Jr., R. L. (2011). *Edison to Enron: Energy Markets and Political Strategies*. John Wiley & Sons.

Cunningham, J. J. (2015). STARS: Manhattan Electric Power Distribution, 1881–1901. *Proceedings of the IEEE*, 103(5), 850–858.

Devine Jr., W. D. (1982). *Historical Perspective on the Value of Electricity in American Manufacturing* (No. ORAU/IEA-82-8 (M)). Oak Ridge Associated Universities, Inc., TN. Inst. for Energy Analysis.

Fairley, P. (2012). DC versus AC: The second war of currents has already begun [in my view]. *IEEE Power and Energy Magazine*, 10(6), 104–103.

Fleming, K. R. (1991). *Power at Cost: Ontario Hydro and Rural Electrification, 1911–1958*. McGill-Queen's Press-MQUP.

Guarnieri, M. (2013). The beginning of electric energy transmission: Part two. *IEEE Industrial Electronics Magazine*, 7(2), 52–59.

Hargadon, A. B., & Douglas, Y. (2001). When innovations meet institutions: Edison and the design of the electric light. *Administrative Science Quarterly*, 46(3), 476–501.

Hughes, T. P. (1993). *Networks of Power: Electrification in Western* Society, *1880–1930*. JHU Press.

Jeszenszky, S. (1996). History of transformers. *IEEE Power Engineering Review*, 16(12), 9.

Kahn, E. (1984). *Crossroads in Electric Utility Planning and Regulation*. American Council for an Energy-Efficient Economy. Washington, DC.

Lambert, J. D. (2015). *The Power Brokers: The Struggle to Shape and Control the Electric Power Industry*. MIT Press.

Martin, T. C. (1911). Frank Julian Sprague. *Scientific American*, *105*(17), 363–364.

McDonald, F. (2004). *Insull: The Rise and Fall of a Billionaire Utility Tycoon*. Beard Books.

Munson, R. (2005). *From Edison to Enron: The Business of Power and What it Means for the Future of Electricity*. Northeast-Midwest Institute.

Nichols, R. S. (2003). The first electric power transmission line in North America-Oregon City, Oregon. *IEEE Industry Applications Magazine*, 9(4), 7–10.

Noberni, R. (1982). Pearl street station – The first truly complete and integrated electric utility system. *IEEE Power Engineering Review*, PER-2(1), 2–3.

Platt, H. L. (1991). *The Electric City: Energy and the Growth of the Chicago Area, 1880–1930*. University of Chicago Press. Retrieved from: https://quod.lib.umich.edu/lib/colllist/ (Accessed 2020 November 17)

Pope, F. L., Phelps, G. M., Martin, T. C., & Wetzler, J. (1887). *The Electrical Engineer: A Weekly Review of Theoretical and Applied Electricity* (Vol. 7). Williams & Company. Retrieved from: https://books.google.ca/books?id%3DuI05AQAAMAAJ%26pg%3DPA383%26dq%3DFig+2+-+%22Shallenberger%27s+Meter%26%23x002B; (case+removed)%22+The+Electrical+Engineer&hl=en&sa=X&redir_esc=y#v=onepage&q=Fig%202%20-%20%22Shallenberger's%20Meter%20(case%20removed)%22%20The%20Electrical%20Engineer&f=false (Accessed 2020 October 8).

Roguin, A. (2004). Nikola Tesla: The man behind the magnetic field unit. *Journal of Magnetic Resonance Imaging: An Official Journal of the International Society for Magnetic Resonance in Medicine*, 19(3), 369–374.

Rosenbloom, D., & Meadowcroft, J. (2014). The journey towards decarbonization: Exploring socio-technical transitions in the electricity sector in the province of Ontario (1885–2013) and potential low-carbon pathways. *Energy Policy*, 65, 670–679.

Ruch, C. A. (1984). George Westinghouse-engineer and DOER!!!. *IEEE Transactions on Industry Applications*, (6), 1395–1402.

Saqib, S. (2019). HVDC: A Building Block for a Resilient, Flexible and Interconnected Grid. *T&D World* Retrieved from: https://www.tdworld.com/overhead-transmission/article/20972318/ (Accessed 2020 September 14)

Skrabec Jr., Q. R., & Skrabec, Q. R. (2007). *George Westinghouse: Gentle Genius*. Algora Publishing.

Smil, V. (2005). *Creating the Twentieth Century: Technical Innovations of 1867–1914 and Their Lasting Impact*. Oxford University Press.

Smil, V. (2010). *Power Density Primer: Understanding the Spatial Dimension of the Unfolding Transition to Renewable Electricity Generation*. Retrieved from: http://vaclavsmil.com/publications/ (Accessed 2021 June 10)

Sowmya, B., Kumar, B. S., & Gangadhar, D. V. (2016). Wireless ARM-based automatic meter reading & control system (WAMRCS). *International Journal of Advanced Technology and Innovative Research*, 8(22),4366–4370.

Sulzberger, C. (2010). Thomas Edison's 1882 pearl street generating station. *IEEE Global History Network*.

Thomas, B. A., Azevedo, I. L., & Morgan, G. (2012). Edison revisited: Should we use DC circuits for lighting in commercial buildings? *Energy Policy*, 45, 399–411.

Tobin, J. (2012). *Great Projects: The Epic Story of the Building of America, from the Taming of the Mississippi to the Invention of the Internet*. Simon and Schuster.

Troesken, W. (2006). Regime Change and Corruption. A History of Public Utility Regulation. In *Corruption and Reform: Lessons from America's Economic History* (pp. 259–282). University of Chicago Press.

Wang, P., Goel, L., Liu, X., & Choo, F. H. (2013). Harmonizing AC and DC: A hybrid AC/DC future grid solution. *IEEE Power and Energy Magazine*, 11(3), 76–83.

Wasik, J. F. (2006). *The Merchant of Power: Sam Insull, Thomas Edison, and the Creation of the Modern Metropolis*. Macmillan.

Chapter 6

Alberta Utilities Commission. (2001, December 12). *Methodology for Managing Gas Supply Portfolios and Determining Gas Cost Recovery Rates*. Proceeding and Gas Rate Unbundling Proceeding. Retrieved from: www.auc.ab.ca/regulatory_documents/ProceedingDocuments/2001/2001–110.pdf (Accessed 2020 November 6)

Bakke, G. (2016). *The Grid: The Fraying Wires Between Americans and our Energy Future*. Bloomsbury Publishing.

Balaraman, K. (2019, November 19). Cuomo Threatens to Revoke National Grid's License to Provide Gas in NYC Due to Hookup Moratorium. *UtilityDive*. Retrieved from: www.utilitydive.com/news/cuomo-national-grid-certificate/567192/ (Accessed 2020 November 18).

Bernstein, M. H. (1955). *Regulating Business by Independent Commission*. Princeton University Press.

Borenstein, S., & Bushnell, J. (2015). *The US Electricity Industry After 20 Years of Restructuring* (No. w21113). National Bureau of Economic Research.

Boyd, W. (2018). Just price, public utility, and the long history of economic regulation in America. *Yale Journal on Regulation*, 35(3), 2.

Bradley Jr., R. L. (1996). The origins of political electricity: Market failure or political opportunism? *Energy Law Journal*, 17(1), 59–102.

Bradley Jr., R. L. (2011). *Edison to Enron: Energy Markets and Political Strategies*. John Wiley & Sons.

Brown, D. P., Eckert, A., & Eckert, H. (2017). Electricity markets in transition: Market distortions associated with retail price controls. *The Electricity Journal*, 30(5), 32–37.

Chastain v. British Columbia Hydro and Power Authority (1972) B.C.J. No. 576; 32 D.L.R. (3d) 443 (BCSC).

Covaleski, M. A., Dirsmith, M. W., & Samuel, S. (1995). The use of accounting information in governmental regulation and public administration: The impact of John R. Commons and early institutional economists. *Accounting Historians Journal*, 22(1), 1–33.

Creamer, D., Dobrovolsky, S. P., Borenstein, I., & Bernstein, M. (1960). *Debt and equity financing. In Capital in Manufacturing and Mining: Its Formation and Financing* (pp. 156–191). Princeton University Press.

Cudahy, R. D., & Henderson, W. D. (2005). From Insull to Enron: Corporate (re) regulation after the rise and fall of two energy icons. *Energy Law Journal*, 26, 35.

Demsetz, H. (1968). Why regulate utilities? *The Journal of Law and Economics*, 11(1),55–65.

Devine Jr., W. D. (1982). *Historical Perspective on the Value of Electricity in American Manufacturing* (No. ORAU/IEA-82-8 (M)). Oak Ridge Associated Universities, Inc., TN. Inst. for Energy Analysis.

Gillis, Justin. (2020, August 8). When utility money talks. *The New York Times*. Retrieved from: www.nytimes.com/2020/08/02/opinion/utility-corruption-energy.html (Accessed2020 November 22)

Ghadessi, M., & Zafar, M. (2017). Utility General Rate Case – A Manual for Regulatory Analysts. *California Public Utilities Commission*, Policy & Planning Division.

Gordon, R. J. (2004). *Two Centuries of Economic Growth: Europe Chasing the American Frontier* (No. w10662). National Bureau of Economic Research.

Hausman, W. J., & Neufeld, J. L. (1989). Engineers and economists: Historical perspectives on the pricing of electricity. *Technology and Culture*, 30(1), 83–104.

Hausman, W. J., & Neufeld, J. L. (2002). The market for capital and the origins of state regulation of electric utilities in the United States. *Journal of Economic History*, 1050–1073.

Hausman, W. J., & Neufeld, J. L. (2004). The economics of electricity networks and the evolution of the US Electric Utility Industry, 1882–1935. *Business and Economic History On-Line*, 2(26).

Hoffman, D. E. (2010). *Scarface Al And the Crime Crusaders: Chicago's Private War Against Capone*. SIU Press.

Insull, S. (1898). Standardization, cost system of rates, and public control. Presidential address to the convention of the National Electric Light Association (June 7). Reprinted in S Insull (1915). *Central-Station Electric Service*, 34–47.

Jacobson, C., Klepper, S., & Tarr, J. A. (1985). Water, electricity and cable television: A study of contrasting historical patterns of ownership and regulation. *FLUX Cahiers Scientifiques Internationaux Réseaux et Territoires*, 1(3), 2–31.

Jonnes, J. (2004). *Empires of light: Edison, Tesla, Westinghouse, and the Race to Electrify the World*. Random House.

Joskow, P. L., & Schmalensee, R. (1986). Incentive regulation for electric utilities. *Yale Journal on Regulation*, 4(1), 2.

Joskow, P. L. (1989). *Regulatory Failure, Regulatory Reform and Structural Change in the Electric Power Industry*. Rev. Draft Jan. 25, 1989.

Knittel, C. R. (2006). The adoption of state electricity regulation: The role of interest groups. *The Journal of Industrial Economics*, 54(2), 201–222.

Lambert, J. D. (2015). *The Power Brokers: The Struggle to Shape and Control the Electric Power Industry*. MIT Press.

Lesser, J. A. (2002). The used and useful test: Implications for a restructured electric industry. *Energy Law Journal*, 23(2), 349.

McDermott, K. (2012). *Cost of Service Regulation in The Investor-Owned Electric Utility Industry*. Edison Electrical Institute.

McDonald, F. (1957). *Let There Be Light: The Electric Utility Industry in Wisconsin, 1881–1955*. American History Research Center.

Melody, W. H. (2002). Designing Utility Regulation for 21st Century Markets. *The Institutionalist Approach to Public Utility Regulation*, Michigan State University Press, East Lansing, MI, pp. 25–81.

Mosca, M. (2008). On the origins of the concept of natural monopoly: Economies of scale and competition. *The European Journal of the History of Economic Thought*, 15(2), 317–353.

Munson, R. (2005). *From Edison to Enron: The Business of Power and What It Means for the Future of Electricity*. Northeast-Midwest Institute.

Ness, E., & Fraley, O. (1957). *The Untouchables*. Messner Publishing, New York.

Nelles, H. V. (2003). Hydro and after: The Canadian experience with the organization, nationalization and deregulation of electrical utilities. *Annales Historiques de l'Electricite*, (1), 117–132.

Nelles, H. (2014). Light switch: Towards a history of the second enlightenment. Scientia Canadensis: *Canadian Journal of the History of Science, Technology and Medicine/Scientia Canadensis: Revue Canadienne d'Histoire des Sciences, des Techniques et de la Médecine*, 37 (1–2), 11–33.

Netherton, A. (2007). The political economy of Canadian hydro-electricity: Between old "provincial hydros" and neoliberal regional energy regimes. *Canadian Political Science Review*, 1(1), 107–124.

Porter, M. E. (1979). How Competitive Forces Shape Strategy. In *Readings in Strategic Management* (pp. 133–143). Palgrave, London.

Smith, V. L. (1996). Regulatory reform in the electric power industry. *Regulation*, 19, 33.

Stevenson, R. L. (1881). A plea for gas lamps. *Virginibus Puerisque*, 249–256. Retrieved from: www.archive.org/stream/virginibuspueris05stev#page/254/mode/2up (Accessed 2020 November 2)

Swartwout, R. L. (1992). Current utility regulatory practice from a historical perspective. *Natural Resources Journal*, 32, 289.

Taylor, A. R. (1962). Losses to the public in the Insull collapse: 1932–1946. *Business History Review*, 36(2), 188–204.

Thakar, N. (2008). The urge to merge: A look at the repeal of the public utility holding company act of 1935. *Lewis & Clark Law Review*, 12, 903.

Tomain, J. P. (1997). Electricity restructuring: A case study in government regulation. *Tulsa Law Journal*, 33, 827.

Troesken, W. (2006). *Regime Change and Corruption. A History of Public Utility Regulation. In Corruption and Reform: Lessons from America's Economic History* (pp. 259–282). University of Chicago Press.

Tuttle, D. P., Gülen, G., Hebner, R., King, C. W., Spence, D. B., Andrade, J., Wible, J., Baldick, R., & Duncan, R. (2016). *The History and Evolution of the US Electricity Industry*. White Paper UTEI/2016-05-2.

Ulmer, M. J. (1960). *Capital in Transportation, Communications, and Public Utilities: Its Formation and Financing*. NBER Books.

US Bureau of Labor Statistics (2020) CPI Inflation Calculator. Retrieved from: https://www.bls.gov/data/inflation_calculator.htm (Accessed 2020 October 28)

Wasik, J. F. (2006). *The Merchant of Power: Sam Insull, Thomas Edison, and the Creation of the Modern Metropolis*. Macmillan.

Introduction to Part 3

DNV GL Energy. (2014). *A Review of Distributed Energy Resources*. Report Prepared by DNV GL Energy for the New York Independent System Operator. Retrieved from: https://www.nyiso.com/documents/20142/3065827/A_Review_of_Distributed_Energy_Resources_September_2014.pdf/ (Accessed 2021 June 5).

US. EIA. (2018f). US Energy Information Administration. Electric Power Monthly with Data for February 2018. US Department of Energy, Washington DC. Retrieved from: https://www.eia.gov/electricity/monthly/(Accessed 2021 January 10)

Chapter 7

Baker, D., & Kaufman, l. (2020, December 10). The Making of Biden's Superfast Push for Clean Electricity. *Bloomberg Green*. Retrieved from: https://www.bloomberg.com/news/features/2020-12-10/how-joe-biden-s-2035-green-energy-grid-could-work (Accessed 2021 January 30)

BloombergNEF. (2018), New Energy Outlook 2018. *Bloomberg New Energy Finance*. Retrieved from: https://about.bnef.com/new-energy-outlook/#toc-download

BloombergNEF. (2020a December 16). Battery Pack Prices Cited Below $100/kWh for the First Time in 2020, While Market Average Sits at $137/kWh. *Bloomberg New Energy Finance*. Retrieved from: https://about.bnef.com/blog/battery-pack-prices-cited-below-100-kwh-for-the-first-time-in-2020-while-market-average-sits-at-137-kwh/ (Accessed 2021 January 24)

BloombergNEF. (2020b). Electric Vehicle Outlook 2020, Executive Summary. *Bloomberg New Energy Finance*. Retrieved from: https://about.bnef.com/electric-vehicle-outlook/ (Accessed 2021 January 22)

Bodis, K., Kougias, I., Jager-Waldau, A., Taylor, N., & Szabo, S. (2019). A high-resolution geospatial assessment of the rooftop solar photovoltaic potential in the European Union. *Renewable and Sustainable Energy Reviews*, 114, 109309.

Bronski, P., Creyts, J., Guccione, L., Madrazo, M., Mandel, J., Rader, B., . . . & Crowdis, M. (2014). The Economics of Grid Defection: When and Where Distributed Solar Generation Plus Storage Competes with Traditional Utility Service. *Rocky Mountain Institute*, 1–73.

Campbell, R. (2018). *The Smart Grid: Status and Outlook*. US Congressional Research Service. Retrieved from: https://fas.org/sgp/crs/misc/R45156.pdf (Accessed 2021 January 30)

Carvallo, J. P., Larsen, P. H., Sanstad, A. H., & Goldman, C. A. (2018). Long term load forecasting accuracy in electric utility integrated resource planning. *Energy Policy*, 119, 410–422.

EurObserv'ER. (2020). Photovoltaic Barometer 2020. EurObserv'er. Retrieved from: https://www.eurobserv-er.org/photovoltaic-barometer-2020/ (Accessed 2021 January 21)

European Commission. (2020). Committing to climate-neutrality by 2050. European Commission. Retrieved from: https://ec.europa.eu/commission/presscorner/detail/en/ip_20_335 (Accessed 2021 January 24)

European Commission. (2021). *Smart Metering Deployment in the European Union*. European Commission Joint Research Centre. Retrieved from: https://ses.jrc.ec.europa.eu/smart-metering-deployment-european-union#:~:text=To%20date%2C%20Member%20States%20have,will%20have%20one%20for%20gas. (Accessed2021 January 30)

Fehrenbacher, K. (2016, October 19). These are the US companies with the most solar power. *Fortune Magazine*. Retrieved from: https://fortune.com/2016/10/19/corporate-solar-target-walmart/ (Accessed 2021 June 13)

Feldman, D., Ramasamy, V., Fu, R., Ramdas, A., Desai, J., & Margolis, R. (2021). *US Solar Photovoltaic System and Energy Storage Cost Benchmark: Q1 2020* (No. NREL/TP-6A20-77324). National Renewable Energy Lab. (NREL), Golden, CO.

Fu, R., Feldman, D., & Margolis, R. (2018). *Solar Photovoltaic System Cost Benchmark: Q1 2018*. National Renewable Energy Lab. (NREL), Golden, CO. Technical Report NREL.TP-6A2-72399 Retrieved from: https://www.nrel.gov/docs/fy19osti/72399.pdf (Accessed 2019 March 2)

Government of Canada. (2018). Greenhouse Gas Emissions and Economic Sectors. Retrieved from: https://www.canada.ca/en/services/environment/weather/climatechange/climate-plan/net-zero-emissions-2050.html (Accessed 2019 February 2)

Government of Canada. (2020). Net-Zero Emissions by 2050. Retrieved from https://www.canada.ca/en/environment-climate-change/services/environmental-indicators/greenhouse-gas-emissions/canadian-economic-sector.html (Accessed 2020 February 2)

Graffy, E., & Kihm, S. (2014). Does disruptive competition mean a death spiral for electric utilities? *Energy Law Review*, *35*, 1–44.

Granovetter, M., & McGuire, P. (1998). The making of an industry: Electricity in the United States. *The Sociological Review*, *46*(S1), 147–173.

Himmelman, J. (2012, August 9). The secret to solar power. *The New York Times*. Retrieved from: https://www.nytimes.com/2012/08/12/magazine/the-secret-to-solar-power.html (Accessed 2021 June 13)

Hirsh, R. F. (2011). Historians of technology in the real world: reflections on the pursuit of policy-oriented history. *Technology and Culture*, *52*(1), 6–20

Holder, M. (2020 December 16). We're getting there: Smart meters on track to reach half of Britain's homes and businesses in 2021. *Business Green*. Retrieved from: https://www.businessgreen.com/feature/4025086/getting-smart-meters-track-reach-half-britain-homes-businesses-2021 (Accessed 2021 January 29)

Howland, E. (2014, February 14). Twelve Unforgettable Quotes from NRG Energy CEO David Crane. *Utility Dive*. Retrieved from: https://www.utilitydive.com (Accessed 2019 February 9)

IEA (International Energy Agency). (2020a). *Global EV Outlook 2020: Entering the Decade of Electric Drive?* International Energy Agency. Retrieved from: https://www.iea.org/reports/ (Accessed 2021 January 29)

Joskow, P. (1989). Regulatory failure, regulatory reform, and structural change in the electrical power industry. *Brookings papers on economic activity. Microeconomics*, 1989, 125–208.

Mallapragada, D. S., Sepulveda, N. A., & Jenkins, J. D. (2020). Long-run system value of battery energy storage in future grids with increasing wind and solar generation. *Applied Energy*, 275, 115390.

National Academies of Sciences, Engineering, and Medicine. 2021. Accelerating Decarbonization of the US Energy System. Washington, DC: The National Academies Press. Retrieved from: https://doi.org/10.17226/25932 (Accessed 2021 January 29)

National Energy Board. (2018). *Canada's Adoption of Renewable Power Sources – Energy Market Analysis*. Government of Canada. Retrieved from: https://www.neb-one.gc.ca/nrg (Accessed 2019 March 5)

Natural Resources Canada. (2018). *About Electricity*. Retrieved from: https://www.nrcan.gc.ca/energy/electricity-infrastructure/about-electricity/7359 (Accessed 2019 March 15)

Re100. (2020). *RE100 Annual Report 2020. The Climate Group*. Retrieved from: http://there100.org/companies

Roselund, C. (2018 April 23), Walmart to host solar power on 130 more sites. *PV Magazine*. Retrieved from: https://pv-magazine-usa.com/2018/04/23/walmart-to-host-solar-power-on-130-more-sites/ (Accessed 2021 June 7)

SEIA. (2020). Solar Industry Research Data. Solar Energy Industries Association. Retrieved from: https://www.seia.org/solar-industry-research-data (Accessed 2020 January 14)

Spector, J. (2020a). Long-Term Value of Grid Storage Is All About Capacity. *GreenTech Media*. Retrieved from: https://www.greentechmedia.com/articles/read/long-term-value-of-grid-storage-is-all-about-capacity-study-finds (Accessed 2021 January 28)

Tomain, J. P. (1997). Electricity Restructuring: A Case Study in Government Regulation. *Tulsa Law Review*, 33, 827.

Tsukimori, O. (2020, November 9). Japan faces high costs in achieving Suga's 2050 carbon neutrality target. *Japan Times*. Retrieved from: https://www.japantimes.co.jp/news/2020/11/09/business/japan-2050-carbon-neutrality-costs/ (Accessed 2021 January 24)

US Bureau of Economic Analysis. (2021, January 24). Real gross domestic product (GDPC1 Dataset). FRED, Federal Reserve Bank of St. Louis. Retrieved from: https://fred.stlouisfed.org/series/GDPC1 (Accessed 2021 January 24)

US EIA. (Energy Information Administration). (2018). *Annual energy outlook 2018*. US Energy Information Administration. Retrieved from: https://www.eia.gov/outlooks/aeo/pdf/AEO2018.pdf (Accessed 10 July 2020)

US EIA. (Energy Information Administration). (2020a). *Annual Energy Outlook 2020: Electricity*. US Energy Information Administration. Retrieved from: https://www.eia.gov/outlooks/aeo/ (Accessed 2021 January 21)

US EIA. (Energy Information Administration). (2020b). *Electricity explained*. US Energy Information Administration. Retrieved from: https://www.eia.gov/outlooks/aeo/ (Accessed 2021 January 29)

US EIA. (Energy Information Administration). (2021a). *Retail sales of electricity to ultimate customers, Table_7.1_Electricity_Overview.xls*. US Energy Information Administration. Retrieved from: https://www.eia.gov/electricity/data.php (Accessed 2021 January 29)

US EIA. (Energy Information Administration). (2021b). *Advanced Metering Count by Technology Type*. US Energy Information Administration. Retrieved from: https://www.eia.gov/electricity/annual/html/epa_10_10.html (Accessed 2021 January 31)

US EIA. (Energy Information Administration). (2021c). *Monthly Energy Review, Table 7.2a, March 2020*. US Energy Information Administration. Retrieved from: https://www.eia.gov/ener gyexplained/electricity/electricity-in-the-us.php (Accessed 2021 January 31)

US EPA. (Environmental Protection Agency). (2021). *Global Greenhouse Gas Emissions Data*. US Environmental Protection Agency. Retrieved from: https://www.epa.gov/ghgemissions/global-greenhouse-gas-emissions-data (Accessed 2021 January 31) Note that US government publications are in the public domain. For usage permissions, see https://www.eia.gov/about/copyrights_reuse.php

Walton, R. (2021, January 12). 2021 Outlook: The future of electric vehicle charging is bidirectional – but the future isn't here yet. *Utility Dive*. Retrieved from: https://www.utilitydive.com/news/2021-outlook-the-future-of-electric-vehicle-charging-is-bidirectional-bu/592957/ (Accessed 2021 January 26)

World Bank Group. (2021a). *Electric Power Consumption (kWh per capita)*. World Bank Group Retrieved from: https://data.worldbank.org/indicator/EG.USE.ELEC.KH.PC?locations=US-CA-JP-EU (Accessed 2021 January 26)

World Bank Group. (2021b). *GDP Per Unit of Energy Use (constant 2017 PPP $ per kg of oil equivalent)*. Retrieved from: https://data.worldbank.org/indicator/EG.GDP.PUSE.KO.PP.KD?lo cations=JP-CA-US-EU (Accessed 2021 January 26)

Chapter 8

Associated Press. (2020, June 16). PG&E confesses to killing 84 people in 2018 California fire as part of guilty plea. *The Guardian Newspaper*. Retrieved from: https://www.theguardian.com/business/2020/jun/16/pge-california-wildfire-camp-fire-paradise-guilty-plea (Accessed 2021 January 4)

Bade, G. (2018, July 9). Heat wave cuts power to over 80K around Los Angeles. *Utility Dive Magazine*. Retrieved from: https://www.utilitydive.com

Baden-Fuller, C., & Mangematin, V. (2013). Business models: A challenging agenda. *Strategic Organization, 11*(4), 418–427.

BBC News. (1998, June 8). *Sport: Football trouble brewing for National Grid*. BBC Online Network Retrieved from: http://news.bbc.co.uk/2/hi/uk_news/109355.stm (Accessed 2020 December 20)

BC Hydro. (2019). *BC Hydro Quick Facts*. Retrieved from: https://www.bchydro.com/content/dam/BCHydro/customer-portal/documents/corporate/accountability-reports/financial-reports/an nual-reports/BCHydro-Quick-Facts-20190331.pdf (Accessed 2020 December 18)

BloombergNEF. (2020). *Electric Vehicle Outlook 2020, Executive Summary*. Bloomberg NEF. Retrieved from: https://about.bnef.com/electric-vehicle-outlook/ (Accessed 2021 January 4)

California Energy Commission. (2020). *2020 Statistics and Charts, NEM Solar PV*. California Distributed Generation Statistics. Retrieved from: https://www.californiadgstats.ca.gov/ (Accessed 2021 January 4)

California ISO. (2013). *Demand Response and Energy Efficiency Roadmap: Maximizing Preferred Resources*. Folsom, CA: California Independent Systems Operator. Retrieved from: http://www.caiso.com/documents/dr-eeroadmap.pdf (Accessed 2021 December 12)

California ISO. (2016). *FastFacts – What the Duck Curve Tells Us About Managing a Green Grid*. California Independent System Operator. Retrieved from: www.caiso.com/Documents/Flexi bleResourcesHelpRenewables_FastFacts.pdf (Accessed 2020 December 18)

California ISO. (2018). *Briefing on Renewables and Recent Grid Operations, March 21- 22,2018.* California Independent System Operator. Retrieved from: https://www.caiso.com/Documents/ Briefing_Renewables_RecentGridOperations-Presentation-Mar2018.pdf (Accessed 2020 December 18)

Cochran, J., Lew, D., & Kumar, N. (2013). *Flexible Coal: Evolution from Baseload to Peaking Plant.* National Renewable Energy Laboratory, Golder.

Cunningham, J. J. (2015). STARS: Manhattan Electric Power Distribution, 1881–1901. *Proceedings of the IEEE*, 103(5), 850–858.

Das, H. S., Rahman, M. M., Li, S., & Tan, C. W. (2020). Electric vehicles standards, charging infrastructure, and impact on grid integration: A technological review. *Renewable and Sustainable Energy Reviews*, 120, 109618.

Denholm, P., Margolis, R., & Milford, J. (2008). *Production cost modeling for high levels of photovoltaics penetration* (No. NREL/TP-581-42305). National Renewable Energy Lab (NREL), Golden, CO.

Denholm, P., O'Connell, M., Brinkman, G., & Jorgenson, J. (2015). *Overgeneration From Solar Energy in California. A Field Guide to the Duck Chart* (No. NREL/TP-6A20-65023). National Renewable Energy Lab (NREL), Golden, CO.

Donadee, J., Shaw, R., Garnett, O., Cutter, E., & Min, L. (2019). Potential benefits of vehicle-to-grid technology in California: High value for capabilities beyond one-way managed charging. *IEEE Electrification Magazine*, 7(2), 40–45.

European Environment Agency (2016). *Electric Vehicles and the Energy Sector – Impacts on Europe's Future Emissions*. EEA Briefing no. 02/2016.

Fox-Penner, P. (2014). *Smart Power Anniversary Edition: Climate Change, The Smart Grid, and the Future of Electric Utilities*. Island Press.

Held, L., Märtz, A., Krohn, D., Wirth, J., Zimmerlin, M., Suriyah, M. R., . . ., & Fichtner, W. (2019). The influence of electric vehicle charging on low voltage grids with characteristics typical for Germany. *World Electric Vehicle Journal*, 10(4), 88.

IEA. (2020). *Global EV Outlook 2020*, International Energy Agency, Paris. Retrieved from: https:// www.iea.org/reports/global-ev-outlook-2020 (Accessed 4 January 2021)

Kim, C. Y., Kim, C. R., Kim, D. K., & Cho, S. H. (2020). Analysis of challenges due to changes in net load curve in South Korea by integrating DERs. *Electronics*, 9(8), 1310.

Kosowatz, J. (2018). Energy storage smooths the duck curve. *Mechanical Engineering*, 140(06), 30–35.

Lyons, C. (2019, June 6). Non-Wires Alternatives: Non-Traditional Solutions to Grid Needs. *T&D World*. Retrieved from: https://www.tdworld.com/overhead-distribution/article/20972703/non wires-alternatives-nontraditional-solutions-to-grid-needs (Accessed 2021 January 4)

Muratori, M. (2018). Impact of uncoordinated plug-in electric vehicle charging on residential power demand. *Nature Energy*, 3(3), 193–201.

Morris, J. D. (2019, January 4). California Fires Add Fuel to the Push for More Solar Energy. *San Francisco Chronicle*. Retrieved from: http://www.govtech.com/fs/infrastructure/California-Fires-Add-Fuel-to-the-Push-for-More-Solar-Energy.html (Accessed 2021 January 15)

NERC. (2020). *2020 Long-Term Reliability Assessment*. North American Electric Reliability Corporation.

NREL. (2018). *Ten Years of Analyzing the Duck Chart*. National Renewable Energy Laboratory Retrieved from: https://www.nrel.gov/news/program/2018/10-years-duck-curve.html (Accessed 2020 December 16)

NYPSC, New York Public Service Commission. (2015). Case 14-M-0101 – Proceeding on Motion of the Commission in Regard to Reforming the Energy Vision. Order Adopting Regulatory Policy Framework and Implementation Plan. February, 26(2015), 11. New York Public Service Commission. Retrieved from: http://www3.dps.ny.gov/

Ramchurn, S. D., Vytelingum, P., Rogers, A., & Jennings, N. R. (2012). Putting the 'Smarts' into the Smart Grid: A Grand Challenge for Artificial Intelligence. *Communications of the Association for Computing Machinery*, *55*(4), 86–97.

Romero, D. (2020, October 23). Power outages planned for nearly 500,000 as California braces for more fires. *NBC News*. Retrieved from: www.nbcnews.com/news/us-news/power-outages-planned-nearly-500-000-california-braces-more-fires-n1244619 (Accessed 2021 January 4)

Roth, S. (2019, December 12). California now has 1 million solar roofs. Are 1 million batteries next? *LA Times*. Retrieved from: https://www.latimes.com/environment/story/2019-12-12/california-clean-energy-milestone-1-million-solar-roofs (Accessed 2021 January 4)

Shellenberger, M. (2018, July 26). As heatwave tests the limits of renewables, anti-nuclear governments return to nuclear. *Forbes Magazine*. Retrieved from: https://www.Forbes.com (Accessed 2020 December 13)

Spector, J. (2018, April 23). Massachusetts Is Staring Down a Duck Curve of Its Own. Storage Could Help. *GreenTech Media*. Retrieved from: https://www.greentechmedia.com/articles/read/massachusetts-is-staring-down-a-duck-curve-of-its-own-storage-could-help (Accessed 2020 December 14)

St. John, J. (2020, September 22). Unlocking California's Gigawatt-Scale Distributed Energy Potential. *GreenTech Media*. Retrieved from: https://www.greentechmedia.com/articles/read/unlocking-californias-gigawatt-scale-distributed-energy-potential (Accessed 2021 January 18)

St. John, J. (2016, November 3). The California Duck Curve Is Real, and Bigger Than Expected. *GreenTech Media*. Retrieved from: https://www.greentechmedia.com/articles/read/the-california-duck-curve-is-real-and-bigger-than-expected (Accessed 2020 December 12)

US EIA. (2020c). *Use of Energy Explained, Energy Use in Homes*. US Energy Information Administration. Retrieved from: https://www.eia.gov/energyexplained/use-of-energy/homes.php (Accessed 2021 January 9)

Walton, R. (2021, January 12). Outlook: The future of electric vehicle charging is bidirectional – but the future isn't here yet. *Utility Dive*. Retrieved from: https://www.utilitydive.com/news/2021-outlook-the-future-of-electric-vehicle-charging-is-bidirectional-bu/592957/ (Accessed 2021 January 26)

Wasik, J. F. (2006). *The Merchant of Power: Sam Insull, Thomas Edison, and the Creation of the Modern Metropolis*. Macmillan.

Wood, E. (2016, November 14). Improving the Scenery: Why Duke's Solar Microgrid is a Very Unusual Energy Project. *Microgrid Knowledge*. Retrieved from: https://microgridknowledge.com/solar-microgrid-duke/ (Accessed 2021 January 2)

Wood Mackenzie. (2020). *Market Design, DERs and the Future of Flexibility: Lessons from California's 2020 Rolling Blackouts*. Wood Mackenzie.

Chapter 9

Baden-Fuller, C., & Mangematin, V. (2013). Business Models: A Challenging Agenda. *Strategic Organization*, *11*(4), 418–427.

Hawaiian Electric. (2020, January 17). *2019 Saw 21% Jump in Solar Generation Capacity*. Hawaiian Electric. Retrieved from: https://www.hawaiianelectric.com/2019-saw-21-percent-jump-in-solar-generation-capacity#~: (Accessed 2021 January 11)

Kennerly, J., & Proudlove, A. (2015). *Going Solar in America: Ranking Solar's Value to Consumer's in America's Largest Cities*. NC State University, Raleigh, NC: NC Clean Energy Technology Center.

Kotler, P. (1986). The Prosumer Movement: A New Challenge for Marketers. *ACR North American Advances*. The Association for Consumer Research.

Metcalfe, M. (2016, June 19). Smart Inverters: Behind-the-Meter Grid Allies. *Power Grid International*. Retrieved from: https://www.power-grid.com/der-grid-edge/smart-inverters-behind-the-meter-grid-allies (Accessed 2021 January 11)

Mulkern, A. (2021, January 4). California Is Closing the Door to Gas in New Homes. E&E News on January 4, 2021. *Scientific American*. Retrieved from: https:http://www.scientificamerican.com. (Accessed 2021 January 11)

Potts, B. (2015 May 17). The hole in the rooftop solar-panel craze. *The Wall Street Journal*. Retrieved from: https://www.wsj.com/articles/the-hole-in-the-rooftop-solar-panel-craze-1431899563 (Accessed 2021 January 11)

Smith, V. L. (1996). Regulatory reform in the electric power industry. *Regulation*. 19, 33.

Toffler, A., & Alvin, T. (1980). *The Third Wave* (Vol. 484). New York: Bantam Books.

Wolsen, M. (2014, January 14). What Google Really Gets Out of Buying Nest for $3.2 Billion, *Wired Magazine*. Retrieved from: https://www.Wired.com (Accessed 2021 January 13)

Chapter 10

Aniti, L. (2017). *Electricity Prices Reflect Rising Delivery Costs, Declining Power Production Costs*. US Energy Information Administration.

Associated Press. (2004, April 13). PG&E Ends 3 Years Under Chapter 11. *Los Angeles Times*. Retrieved from: https://www.latimes.com/archives/la-xpm-2004-apr-13-fi-pge13-story.html (Accessed 2021 January 7)

Averch, H., & Johnson, L. L. (1962). Behavior of the firm under regulatory constraint. *The American Economic Review*, 1052–1069.

Becker, T., & Polson, J. (2005, December 21). Hobbled Calpine Files for Chapter 11. *The Washington Post*. Retrieved from: https://www.washingtonpost.com/archive/business/2005/12/21/hobbled-calpine-files-for-chapter-11/293d0290-b22c-4292-9e13-d18c4053bd34/ (Accessed 2021 January 7)

Berkshire Hathaway. (2015). *2015 Annual Report, Berkshire Hathaway Inc.* Retrieved from: http://www.berkshirehathaway.com (Accessed 2021 January 7)

Berry, J. (1988, January 19). Public Service of N.H. Files for Chapter 11. *The Washington Post*. Retrieved from: https://www.washingtonpost.com/archive/business/1988/01/29/public-service-of-nh-files-for-chapter-11/891cd39e-c273-4458-9a76-2697a87c27b3/ (Accessed 2021 January 7)

Bird, L., McLaren, J., Heeter, J., Linvill, C., Shenot, J., Sedano, R., & Migden-Ostrander, J. (2013). *Regulatory considerations associated with the expanded adoption of distributed solar* (No. NREL/TP-6A20-60613). National Renewable Energy Laboratory (NREL), Golden, CO.

Bloomberg. (2001, December 16). The Fall of Enron. *Bloomberg Business Week* Retrieved from: https://www.bloomberg.com/news/articles/2001-12-16/the-fall-of-enron (Accessed 2021 January 7)

CPUC, The California Public Utilities Commission. (2013). *California Net Energy Metering Ratepayer Impacts Evaluation*. California Public Utilities Commission.

CPUC, The California Public Utilities Commission. (2017). *Workshop on Integrated Distributed Energy Resources July 10, 2017*. The California Public Utilities Commission. Retrieved from: http://www.cpuc.ca.gov (Accessed 2020 January 7)

EIA. (2020, December 15). *Average monthly electricity bill for US residential customers declined in 2019*. US Energy Information Agency. Retrieved from: https://www.eia.gov/todayinenergy/detail.php?id=46276 (Accessed 2021 January 7)

Eisen, J. B. (2013). An Open Access Distribution Tariff: Removing Barriers to Innovation on the Smart Grid. *UCLA Law Review, 61*, 1712.

Faruqui, A., & Bourbonnais, C. (2020). Time of Use Rates: An International Perspective. *Energy Regulation Quarterly*, 8 (2). Retrieved from: https://www.energyregulationquarterly.ca/articles/time-of-use-rates-an-international-perspectives#sthash.yO8wcmUR.pRfZbn9y.dpbs (Accessed 2021 January 14)

Flessner, D. (2018, September 6). Environmental groups sue TVA to overturn new grid access charge. *Chattanooga Times Free Press*. Retrieved from: https://www.timesfreepress.com (Accessed 2020 January 7)

Gilliland, T., & Teufel, A. (2011). *Fisher Investments on Utilities* (Vol. 28). John Wiley & Sons.

Lazar, J., Weston, F., & Shirley, W. (2016). *Revenue regulation and decoupling: A guide to theory and application*. Regulatory Assistance Project.

NAPEE. (2007). *Aligning Utility Incentives with Investment in Energy Efficiency. National Action Plan for Energy Efficiency*. National Action Plan for Energy Efficiency. Prepared by Val R. Jensen, ICF International. Retrieved from: https://www.epa.gov/sites/production/files/2015-08/documents/incentives.pdf (Accessed 2021 January 14)

NREL. (2009). *Decoupling Policies: Options to Encourage Energy Efficiency Policies for Utilities*. National Renewable Energy Laboratory Retrieved from: https://www.energy.gov/eere/downloads/decoupling-policies-options-encourage-energy-efficiency-policies-utilities-clean (Accessed 2021 January 17)

Ralff-Douglas, K., & Zafar, M. (2015). *Electric Utility Business and Regulatory Models*. California Public Utilities Commission, Policy & Planning Division. Retrieved from: http://www.cpuc.ca.gov/ (Accessed 2021 January 5)

Randazzo, R. (2018, March 5). SRP settlement with Tesla could make solar, batteries more affordable. *The Arizona Republic*. Retrieved from: https://www.azcentral.com/story/money/business/energy/2018/03/05/srp-settlement-tesla-could-make-solar-batteries-more-affordable/396385002/ (Accessed 2021 January 7)

Spiegel-Feld, D., & Mandel, B. (2015). *Reforming Electricity Regulation in New York State: Lessons from the United* Kingdom. *Roundtable Report*. Guarini Center, NYU Law. Retrieved from: https://guarinicenter.org/wp-content/uploads/2015/01/RIIO-Roundtable-Report1.pdf (Accessed 2021 January 7)

US EIA. (2020c). *Use of Energy Explained, Energy Use in Homes*. US Energy Information Administration. Retrieved from: https://www.eia.gov/energyexplained/use-of-energy/homes.php (Accessed 2021 January 9)

Wasik, J. F. (2006). *The Merchant of Power: Sam Insull, Thomas Edison, and the Creation of the Modern Metropolis*. Macmillan.

Wood, L., Hemphill, R., Howat, J., Cavanagh, R., Borenstein, S., Deason, J., . . ., & Schwartz, L. (2016). *Recovery of Utility Fixed Costs: Utility, Consumer, Environmental and Economist Perspectives* (No. LBNL-1005742). Lawrence Berkeley National Lab. (LBNL), Berkeley, CA.

Yahoo Finance. (2020, May 5). *Warren Buffett and Greg Abel discuss Berkshire Hathaway's capital expenditures*. Yahoo Finance. Retrieved from: www.youtube.com/watch?v=oZ3xmBxt8zU (Quotation at 4:30 mark). (Accessed 2021 January 15).

Chapter 11

Alizon, F., Shooter, S. B., & Simpson, T. W. (2009). Henry Ford and the Model T: lessons for product platforming and mass customization. *Design Studies*, 30(5), 588–605.

Baden-Fuller, C., & Mangematin, V. (2013). Business models: A challenging agenda. *Strategic Organization*, 11(4), 418–427.

Bakke, G. (2016). *The Grid: The Fraying Wires Between Americans and our Energy Future*. Bloomsbury Publishing.

Barbose, G., & Satchwell, A. J. (2020). Benefits and costs of a utility-ownership business model for residential rooftop solar photovoltaics. *Nature Energy*, 5(10), 750–758.

Blansfield, J., Wood, L., Katofsky, R., Stafford, B., Waggoner, D., & Schwartz, L. C. (2017). *Value-Added Electricity Services: New Roles for Utilities and Third-Party Providers*. Berkeley, CA: Lawrence Berkeley National Laboratory.

Briscoe, B., Odlyzko, A., & Tilly, B. (2006). Metcalfe's law is wrong-communications networks increase in value as they add members-but by how much? *IEEE Spectrum*, 43(7), 34–39.

Brown, B., & Anthony, S. D. (2011). How P&G tripled its innovation success rate. *Harvard Business Review*, 89(6), 64–72.

Bussewitz, C., & Krisher, T (2021 January 21). Biden's climate steps could have big impact on energy firms. *Associated Press*. Retrieved from: https://apnews.com/article/joe-biden-donald-trump-technology-public-health-climate-f8ba1a8e7982227fd27f492f22d771b4 (Accessed 2021 January 22)

Cross-Call, D., Gold, R., Guccione, L., Henchen, M., & Lacy, V. (2018). *Reimagining the Utility: Evolving the Functions and Business Model of Utilities to Achieve a Low-Carbon Grid*. Rocky Mountain Institute.

Curry, B., & Giovannetti, J. (2018, July 18). Ontario Premier Doug Ford courts allies in carbon fight with Ottawa. *The Globe and Mail*. Retrieved from: https://www.theglobeandmail.com (Accessed 2019 January 22)

Kassakian, J., et al., (2011). *The Future of the Electric Grid: An Interdisciplinary MIT Study*. Cambridge, MA: MIT Press. Retrieved from: https://energy.mit.edu/ (Accessed 2020 January 18)

Manshreck, J. (2019). *Technology and Business Model Transformation in Mature Industries, A Study of Business Model Change by Electric Utilities Facing Technological Disruption* (Doctoral dissertation, Grenoble Ecole de Management, Grenoble, France).

Massa, L., Tucci, C., & Afuah, A. (2016). A Critical Assessment of Business Model Research. *Academy of Management Annals*, Annals-2014.

Mirafzal, B., & Adib, A. (2020). On grid-interactive smart inverters: Features and advancements. *IEEE Access*, 8, 160526–160536.

Motyka, M., Thomson, J., Hardin, K., & Sanborn, S. (2020). *Deloitte Resources Study 2020 Energy Management: Paused by pandemic but poised to prevail*. Technical Report. Retrieved from: https://www2.deloitte.com/content/dam/insights/us/articles/6655_Resources-study-2020/DI_Resources-study-2020.pdf (Accessed 2021 January 20)

NCSL. (2020). *State Renewable Portfolio Standards and Goals*. National Conference of State Legislatures.

Pérez-Arriaga, I., & Knittle, C. (2016). *Utility of the future: An MIT Energy Initiative Response to An Industry in Transition*. MIT Energy Initiative.

Porter, M. E. (1979). How Competitive Forces Shape Strategy. In *Readings in Strategic Management* (pp. 133–143). Palgrave, London.

Rathi, A. (2021, January 5). Climate Action is Embedding into How the World Works. *Bloomberg*. Retrieved from: https://www.bloomberg.com/news/articles/2021-01-05/climate-action-is-embedding-into-how-the-world-works (Accessed 2021 January 23)

Satchwell, A., & Cappers, P. (2018). *Evolving Grid Services, Products, and Market Opportunities for Regulated Electric Utilities*. Lawrence Berkeley National Laboratory, Berkeley, CA.

Smith, V. L. (1996). Regulatory reform in the electric power industry. *Regulation*, 19, 33.

Tiernan, T. (2014, November 6). Electric utility 'death spiral' in US is premature, Moody's Says. *S&P Global Platts*. Retrieved from: https://www.spglobal.com/platts/en (Accessed 2019 August 15)

Wood, L., & Borlick, R. (2013). Value of the Grid to DG Customers. *Institute for Electric Innovation Issue Brief*.

Zhang, X. Z., Liu, J. J., & Xu, Z. W. (2015). Tencent and Facebook data validate Metcalfe's law. *Journal of Computer Science and Technology*, 30(2), 246–251.

Zott, C., & Raphael A. (2010). Business model design: An activity system perspective. *Long Range Planning*, 43(2), 216–226.

Chapter 12

Abel, K., Agbim, C., et al. (2017). *A Comparison of New Electric Utility Business Models*. The University of Texas at Austin.

Bakke, G. (2016). *The Grid: The Fraying Wires Between Americans and Our Energy Future*. Bloomsbury Publishing.

Baker, D., & Kaufman, l. (2020, December 10). The Making of Biden's Superfast Push for Clean Electricity. *Bloomberg Green*. Retrieved from: https://www.bloomberg.com/news/features/2020-12-10/how-joe-biden-s-2035-green-energy-grid-could-work (Accessed 2021 January 30)

Barbose, G., & Satchwell, A. J. (2020). Benefits and costs of a utility-ownership business model for residential rooftop solar photovoltaics. *Nature Energy*, 5(10), 750–758.

Blumberg, J., & Luke J. (2009). Wireless-Only and Wireless-Mostly Households: A growing challenge for telephone surveys. National Center for Health Statistics. Retrieved from: http://www.shadac.org/sites/default/files/Old_files/WorkshopSess1-1_Blumberg.pdf (Accessed 2021 February 4)

Blumberg, J., & Luke J (2020). Wireless Substitution: Early Release of Estimates from the National Health Interview Survey, January-June 2020. National Center for Health Statistics.

Comin, D., Hobijn, B., & Rovito, E. (2006). *Five Facts You Need to Know About Technology Diffusion* (No. w11928). National Bureau of Economic Research.

Corneli, S., Kihm, S., & Schwartz, L. (2015). *Electric Industry Structure and Regulatory Responses in a High Distributed Energy Resources Future* (No. LBNL-1003823). Lawrence Berkeley National Lab. (LBNL), Berkeley, CA.

Cross-Call, D., Goldenberg, C., Guccione, L., Gold, R., & O'Boyle, M. (2018). *Navigating Utility Business Model Reform: A Practical Guide to Regulatory Design*. Rocky Mountain Institute.

Cunningham, J. J. (2015). STARS: Manhattan electric power distribution, 1881–1901. *Proceedings of the IEEE*, 103(5), 850–858.

Dyer, T. C., & Martin, F. L. (1910). *Edison, His Life and Inventions*. New York: Harper & Brothers. Retrieved from: http://gutenberg.org/files/820/820-h/820-h.htm (Accessed 2020 November 15)

Eurostat. (2021). Electricity price statistics. European Commission. Retrieved from: https://ec.europa.eu/eurostat/statistics-explained/index.php/Electricity_price_statistics (Accessed 2021 February 11)

Garud, R., & Karnoe, P. (2001). Path creation as a process of mindful deviation. *Path Dependence and Creation, 138.*

Gilliland, T., & Teufel, A (2011). *Fisher Investments on Utilities* (Vol. 28). John Wiley & Sons.

Graffy, E., & Kihm, S. (2014). Does Disruptive Competition Mean a Death Spiral for Electric Utilities? *Energy Law Review, 35,* 1.

Granovetter, M., & McGuire, P. (1998). The Making of an Industry: Electricity in the United States. *The Sociological Review, 46*(S1), 147–173.

Hughes, T. P. (1993). *Networks of Power: Electrification in Western* Society, *1880–1930.* JHU Press.

Hydro Quebec. (2017). Understanding Québec Hydropower. Retrieved from: https://www.hydroquebec.com/data/developpement-durable/pdf/ghg-emissions.pdf (Accessed 2021 February 11)

Hydro Quebec. (2020). *Comparison of Electricity Prices in Major North American* Cities. *Rates in effect April 1, 2020.* Retrieved from: https://www.hydroquebec.com/data/documents-donnees/pdf/comparison-electricity-prices.pdf (Accessed 2021 February 11)

Kind, P. (2013). *Disruptive Challenges: Financial Implications and Strategic Responses to a Changing Retail Electric Business.* Edison Electric Institute.

Klepper, S., & Thompson, P. (2006). Submarkets and the evolution of market structure. *The RAND Journal of Economics, 37*(4), 861–886.

Lambert, F. (2020, August 10). Tesla is considering a home 'energy package' with solar, Powerwall, EV charger bundle. *Electrek.* Retrieved from: https://electrek.co/2020/08/10/tesla-home-energy-package-solar-powerwall-ev-charger-bundle/ (Accessed 2021 April 4)

Lucchesi, N. (2016, October 28). Elon Musk Plays Solar Salesman. Inverse, Bustle Digital Group. Retrieved from: https://www.inverse.com/article/22951-elon-musk-tesla-solar-city%26%23x003F; (Accessed 2021 April 4)

Mandel, B. (2014). *A primer on utility regulation in the United Kingdom: Origins, aims, and mechanics of the RIIO model.* New York University Law Guarini Center. Retrieved from: http://guarinicenter.org/ (Accessed 2021 February 11)

Melton, R. (2015). *Pacific northwest smart grid demonstration project technology performance report volume 1: Technology performance* (No. PNW-SGDP-TPR-Vol. 1-Rev. 1.0; PNWD-4438, Volume 1). Pacific Northwest National Lab. (PNNL), Richland, WA.

NS Energy. (2020 April 02). Top five US states for wind power production profiled. *NS Energy.* Retrieved from: https://www.nsenergybusiness.com/features/wind-power-states-us/ (Accessed 2021 February 8)

NYPSC (New York Public Service Commission). (2014). *Reforming the Energy Vision: NYS Department of Public Service Staff Report and Proposal.* Case 14-M-0101, April 24.

OG&E (Oklahoma Gas and Electric). (2020). Oklahoma Gas and Electric 2019 Annual Report. https://ogeenergy.gcs-web.com/static-files/fb27b4c9-6f19-4a81-b177-7e37e24b782e

Perez-Arriaga, I. J., Jenkins, J. D., & Batlle, C. (2017). A regulatory framework for an evolving electricity sector: Highlights of the MIT utility of the future study. *Economics of Energy & Environmental Policy, 6*(1), 71–92.

Pyper, J. (2015 May 29). Inside the Minds of Regulators: How Different States are Dealing with Distributed Energy. *Greentech Media.* Retrieved from: https://www.greentechmedia.com/articles/read/what-are-the-most-pressing-issues-facing-public-utility-commissioners (Accessed 2020 February 11)

Ralff-Douglas, K., & Zafar, M. (2015). *Electric Utility Business and Regulatory Models.* California Public Utilities Commission, Policy & Planning Division Retrieved from: http://www.cpuc.ca.gov/ (Accessed 2021 February 11)

Robinson, G. O. (1988). The Titanic remembered: AT&T and the changing world of telecommunications. *Yale Journal on Regulation, 5*(2), 11.

Satchwell, A., & Cappers, P. (2018). *Evolving Grid Services, Products, and Market Opportunities for Regulated Electric Utilities*. Lawrence Berkeley National Laboratory, Berkeley, CA.

Spector, J. (2020b October 16). From Pilot to Permanent: Green Mountain Power's Home Battery Network Is Here to Stay. *GreenTech Media*. Retrieved from: https://www.greentechmedia.com/articles/read/from-pilot-to-permanent-green-mountain-powers-home-battery-network-is-sticking-around (Accessed 2021 January 28)

Spiegel-Feld, D., & Mandel, B. (2015). *Reforming Electricity Regulation in New York State: Lessons from the United* Kingdom. *Roundtable Report*. Guarini Center, NYU Law. Retrieved from: https://guarinicenter.org/wp-content/uploads/2015/01/RIIO-Roundtable-Report1.pdf (Accessed 2021 January 7)

Suarez, F., & Lanzolla, G. (2005). The half-truth of first-mover advantage. *Harvard Business Review*, 83(4), 121–7.

Suarez, F., & Lanzolla, G. (2007). The role of environmental dynamics in building a first mover advantage theory. *Academy of Management Review*, 32(2), 377–392.

Zhao, Y., Von Delft, S., Morgan-Thomas, A., & Buck, T. (2020). The evolution of platform business models: Exploring competitive battles in the world of platforms. *Long Range Planning*, 53(4), 101892.

Glossary

ABB Group. (2018). Glossary of Technical Terms. Retrieved from: https://new.abb.com/glossary (Accessed 2019 October 2)

Bird, L., Hurlbut, D., Donohoo, P., Cory, K., & Kreycik, C. (2009). Examination of the Regional Supply and Demand Balance for Renewable Electricity in the United States through 2015. National Research Energy Laboratory Technical Report NREL/TP-6A2-45041.

DNV GL Energy (2014) *A Review of Distributed Energy Resources*. Report Prepared by DNV GL Energy for the New York Independent System Operator. Retrieved from: https://www.nyiso.com/documents/20142/3065827/A_Review_of_Distributed_Energy_Resources_September_2014.pdf/ (Accessed 2021 June 5).

EIS Council. (2021 January 28). Black Sky Hazards. The Electric Infrastructure Security (EIS) Council. Retrieved from: https://www.eiscouncil.org/BlackSky.aspx (Accessed 2021 January 28)

Federal Energy Regulatory Commission. (2017). Assessment of Demand Response and Advanced Metering Staff Report, December 2017. *Federal Energy Regulatory Commission*.

Ghadessi, M., & Zafar, M. (2017). *Utility General Rate Case – A Manual for Regulatory Analysts*. California Public Utilities Commission, Policy & Planning Division.

Statistics Canada. (2007). *Statistics Canada, Households and the Environment: Household energy use, by fuel type and by province, 2007 – Average energy use*. Retrieved from: https://www.statcan.gc.ca/pub/11-526-s/2010001/t004-eng.htm (Accessed 2021 June 05.)

US Department of Energy. (2011). Microgrid workshop report. [Online]. Retrieved from: http://energy.gov/sites/prod/files/Microgrid%20Workshop%20Report%20August%202011.Pdf (Accessed 2019 October 2)

US Department of Energy. (2018). *US Energy Information Administration. Electric Power Monthly with Data for February 2018*. US Department of Energy, Washington DC. Retrieved from: https://www.eia.gov/electricity/monthly/epm_table_grapher.php?t=epmt_6_07_b (Accessed 2020 January 24)

US Department of Energy. (2018f). *US Energy Information Administration. Electric Power Monthly with Data for February 2018*. US Department of Energy, Washington DC. Retrieved from: https://www.eia.gov/electricity/monthly/ (Accessed 2019 December 14)

U.S. Energy Information Administration. (2018b). *How much electricity does an American home use?* [Online] Retrieved from: https://www.eia.gov/tools/faqs/faq.php?id=97&t=3 (Accessed 2020 October 11)

List of Figures

https://doi.org/10.1515/9783110714036-019

List of Tables

https://doi.org/10.1515/9783110714036-020

About the Author

John Manshreck holds a master's degree from the London Business School and a Doctorate of Business Administration from the Grenoble Ecole de Management. John's research has focused on the transformation of business models in mature organizations, particularly those impacted by today's drive to decarbonize the economy. He is also an experienced consultant to the utility and resource sectors, and has been part of many organizational transformations. He is a CPA and has specialist qualifications in project management and information technology. John lives with his family in North Vancouver, British Columbia.

https://doi.org/10.1515/9783110714036-021

Index

https://doi.org/10.1515/9783110714036-022